Outlook of the North East India

About the Authors

Kamal Kumar Datta an Economics honours graduate, post graduated from Calcutta University and did his PhD in Economics from Kurukshetra University and Masters in Business Management from Annamalai University. In 2010, he was awarded ICAR Award for Outstanding Interdisciplinary Team Research 2007-08 in Social Sciences. Presently he is working on sustainable agricultural system to develop a suitable farming system alternative to the existing intensive/extensive method of cultivation. He has more than 100 research publications in various peer-reviewed and refereed research journals of national and international repute. He has contributed to more than 20 technical bulletins and other publications. He has received several awards for best papers as well as best presentation in well reputed leading journals like *Indian Journal of Agricultural Economics* and *Agricultural Economics Research Association*.

Subhasis Mandal, after graduating from Bidhan Chandra Krishi Viswavidayalaya (BCKVV), Mohanpur (West Bengal) completed his Masters in Agricultural Economics from Govind Ballabh Pant University of Agriculture and Technology (GBPUA&T), Pantnagar (Uttarakhand) and PhD in Agricultural Economics from Indian Agricultural Research Institute (IARI), New Delhi. In 2005, he received Dr. D.T. Doshi Award for best paper presentation by Agricultural Economics Research Association (AERA). As team member, he has been recognised and received 'Ground Water Augmentation Award 2008' under the Institution category in the northeast region, bestowed by Ministry of Water Resources, Government of India. He has more than 20 research publications in various peer-reviewed and refereed research journals of national and international repute and has contributed to several technical bulletins, books and other publications. Since 2008, he is working as Senior Scientist (Agricultural Economics), at Central Soil Salinity Research Institute (CSSRI), Regional Research Station (RRS), Canning Town, West Bengal.

Outlook of the North East India

An Agricultural Perspective

KAMAL KUMAR DATTA
SUBHASIS MANDAL

ACADEMIC FOUNDATION
NEW DELHI

www.academicfoundation.com

First published in 2011
by

ACADEMIC FOUNDATION
4772-73 / 23 Bharat Ram Road, (23 Ansari Road),
Darya Ganj, New Delhi - 110 002 (India).
Phones : 23245001 / 02 / 03 / 04.
Fax : +91-11-23245005.
E-mail : books@academicfoundation.com
www.academicfoundation.com

Cataloging in Publication Data--DK
Courtesy: D.K. Agencies (P) Ltd. <docinfo@dkagencies.com>

Datta, Kamal Kumar.
Outlook of the North East India : an agricultural perspective / Kamal Kumar Datta, Subhasis Mandal.
p. cm.
Includes bibliographical references (p.).
ISBN 13: 9788171888962
ISBN 10: 8171888968

1. Agriculture--Economic aspects--India, Northeastern. 2. Economic development--India, Northeastern. 3. Rural development--India, Northeastern. 4. Agriculture--India, Northeastern--Finance. I. Mandal, Subhasis, joint author. II. Title.

DDC 338.109541 22

Typeset by Italics India, New Delhi.
Printed and bound in India.

Dedicated to our beloved Snigdha

Contents

Preface

The northeast region of India is historically, socially, culturally, and politically very complex and simultaneously contains a great diversity of environmental and natural resource base. The region has great potential for larger contribution to the country in many sectors including agriculture. Continuous efforts are being made for all-round development of this region where agriculture must play a very proactive and indispensable role. Some signatures of success stories are being already seen within agriculture, like development of horticultural high value crops. The region needs leapfrog in terms of accelerating agricultural growth and development of the people dependent on it. Currently, the region is facing many dichotomous challenges such as how to accelerate the pace of development without affecting the natural resource base or how to increase the food production of the region while preserving the thousands of landraces having great value for the society for future varietal development programmes.

The region is abundantly endowed with natural resources like soil, water and suitable climatic conditions for growing many high value crops and also off-season fruits and vegetables. Exploration and utilisation of these underlying potentials may be greatly helpful to combat production and price instability not only for the region but also for the country as a whole. The region needs guidance and policy direction for harnessing the development potential of the agriculture sector. While doing so, experience gained and lesson learned from case studies based on secondary and primary information may be immensely helpful for appropriate decision-making for the region. 'Outlook of the North East India: An Agricultural Perspective' is our sincere initiative to this endeavour. This book will be beneficial to policymakers, academicians, students and all others who are involved in development process of the northeast region of India.

K.K. Datta
Subhasis Mandal

1 Introduction

In recent past, Indian economy has recorded spectacular growth and in fact is one among the fastest growing economies of the world. However like other countries around the world, this economic growth is not always ensuring the balanced economic development in terms of reducing regional imbalances in some parts of our country. The northeast region is the most prominent in this list, which comprises a bunch of less developed states with poor economic performance. The northeast region (NER) of India comprising the states of Arunachal Pradesh, Assam, Manipur, Meghalaya, Mizoram, Nagaland, Sikkim and Tripura, representing a distinct agroclimatic zone of the country. Since independence, continuous efforts have been put forth and in particular during preceding decade several aggressive policy initiatives (e.g. spending 10 per cent of the budget for every ministry) has been taken up for all-round development of these states. Though there are some success stories in few fronts (e.g., high literacy rate, nutritional status among children is higher than national average, etc.), but overall the region still has to bear the stigma of 'outlays without outcome'. Several reasons can be attributed to this economic laggardness, such as, difficult geographical terrain, poor infrastructure, and absence of roads hampering, improper perspective planning failing to have a holistic roadmap covering the various sectors, problem of leakages in most of the developmental projects due to lack of non-imposition of strict accountability and finally, failure to ensure the synergies between developmental goals and efforts with security needs, which must go hand in hand to achieve the developmental targets. The irony of development is that, on one side the region is endowed with enormous natural resource base, such as, fertile soil, plenty of water resources and vegetation with huge diversity but simultaneously the region is housing 32 per cent of its population below poverty line (1999-2000). More than 80 per cent of total population is dependent on agriculture and contributing nearly one-fourth of state GDP, whereas at national level, the same level of contribution (22 per cent, precisely) is coming from a population of 54 per cent only. The NER of India has a total geographical area of 25 million hectares and a population of 9.5 million. The region is characterised by slow growth with deficiency in food grains production, lower agricultural productivity, heavy soil erosion and severe ecological imbalance. The region occupies nearly 8 per cent of total geographical area and supports more than 3.8 per cent of the country's population as per the 2001 census. Agriculture is the mainstay of this region, providing livelihood to majority of the people. But the farming practices in NER are predominated by subsistence farming and shifting cultivation still remains important. The sectoral transformation is occurring at a very slow pace, where the agriculture sector is playing passive and supportive role in the process of economic development.

Agriculture and allied activities is the major employment provider for the workers in this region, which is accounted for nearly 50-70 per cent of the total workforce across the various states. However, the contribution of agricultural output is quite low, ranging in between 27-37 per cent only, which is almost half of the share of labour force, engaged in this sector, implying the low level of labour productivity. Overall the NER is characterised by various problems which are either manmade or naturally occurring. This has resulted in significant decline in forest cover resulting in loss of biodiversity, serious soil erosion problems leading to depletion of plant nutrients, gradual degradation and decline in land productivity and its carrying capacity, silting of major river basins causing recurrent floods, drying up of perennial streams as well as ecological imbalance. Gradual degradation of these production resources has become a problem of major concern and calls for location specific measures to conserve, utilise and manage these resources for optimising production on sustained basis without adversely impairing its quality. Subsistence agriculture is an uncertain activity and therefore, risky, particularly when survival is at stake, and is the dominant factor in retaining the traditional farm practices even in the face of opportunities. The poor farmers are naturally risk averters and prefer to be safe than sorry; they tend to prefer an inferior outcome that is relatively certain to the prospect of a higher average return with a greater degree of risk attached (Thirwall, 1999). Shifting cultivation, locally known as *jhuming* is a dominant system of farming in the region. Nearly 17.87 lakh hectares of land is cultivated annually and more than

4.4 lakh families are practising this traditional system of shifting cultivation (Government of India, 1983). Shifting cultivation in its more traditional and cultural integrated form, is ecologically and economically viable system of agriculture as long as population densities are low and *jhum* cycles are long enough to maintain soil fertility (Anon, 1992). However, due to heavy increasing population pressure on land, low carrying capacity and scarcity of forest land suitable for *jhuming*, the *jhum* cycle have come down from about 20 years ago to an average of 4-5 years at present (Ramakrishna *et al.*, 1982).

Despite these difficulties, the region has enormous potentials due to its rich endowment with natural wealth and locational advantages. Perceived weakness of this region can be converted into strength through proper planning and implementation of the developmental projects. The abundance of natural resources in the northeast makes it an extremely profitable business and investment destination. There is no shortage of opportunities, but vision and imagination is needed to turn these opportunities into reality (Mohan, 2003). Income level of rural families can be substantially enhanced particularly on a mass scale by harnessing the vast readily available potential in the field of agriculture, horticulture, medicinal and aromatic herbs and bamboo, water, wind power and minerals, to name just a few sectors. To cater the economic development, the agriculture sector must be strengthened or reoriented to play much more active and indispensable role in any overall strategy of economic progress. Guiding the agricultural policy, it is essential to assess the performance, contribution and role of the agriculture sector critically. In view of these issues the present policy has been envisaged with the broad objectives as follows:

1. Performance of various sectors in contributing to state income and critical assessment of agricultural development status of the region.
2. Identification of various issues impeding enhancement of agricultural income and prioritisation of activities for higher agricultural growth, and
3. Exploring scope and opportunities of new areas of investment to increase farm income as well as to attract private investment in the agricultural sector.

2 Data and Methodology

Sources of Data

Both secondary as well as primary information has been used for the present study. Secondary information was collected from various sources, such as, Central Statistical Organisation (CSO), Reserve Bank of India (RBI), *Statistical Abstract* (various issues), *Agricultural Statistics at a Glance* (2004), *Basic Statistics of North Eastern Region* (various issues).

Primary data was collected through various socioeconomic surveys covering all northeastern states conducted by Division of Agricultural Economics & Statistics, Indian Council of Agricultural Research (ICAR) Complex for NER Region during the period of 2003-04 to 2005-06. The socioeconomic survey was focussed on various sources of income and consumption pattern for management of rural livelihood. For decomposition of farm income analysis, primary data was collected from households of Umroi and Shroewblei village, which comes under Umroi Block of Ri-Bhoi District of Meghalaya. Annual household income along with different sources was collected during the year 2003-04. The major income sources include income from agriculture, livestock, business activities services and others (contractual work, occasional construction work etc.). The agricultural practices in the villages in Meghalaya are predominantly traditional in nature. The sample farmers were selected randomly from the villages who are actively engaged in cultivation of their land in a farming system mode. Since the entire NER is depicted as sloppy, trace and valley land, the primary database which are used in this report is mainly representative for the entire region.

Analytical Framework

To capture the growth pattern of the agricultural sector as such and rural economy as a whole, valuation is essential in order to capture its farming sector mode of contribution to economy rather than conventional method of cropping base production system. In the NER the contribution of individual crops in physical terms are so insignificant that it cannot reflect the real growth process when we compared with any other states or at national level. This region should be looked upon system base approach where sustainable livelihood pattern is the guiding force rather than market-driven surplus generation through extensive agricultural practices. The goal of agricultural development should be achieved through improvement of farming systems rather not to focus on monocropping following intensive agricultural practices. Agricultural intervention should be made in such a way so to protect the natural resources (such as biodiversity and contribution to ecology) of the region. The pace of agricultural development should not be biased to higher growth only at the cost of degrading the unique ecology of the region.

Extent of diversification is important to assess the agricultural performance of the NER. Simpson index has been used to understand the extent and magnitude of diversification. The index value implies that as the magnitude will increase the diversification also increases and value ranges from 0-1. Analysis is based on cross-section data pertaining to three discrete years—1990-91, 1995-96 and 2000-01 to capture the change in diversification pattern over the period. Simple statistical tool such as percentage and compound growth has been employed to analyse the secondary data. Annual compound growth rate (ACGR) has been estimated for various parameters following standard compound growth estimation method.

Gini coefficient has been estimated on sample farmers to derive useful interpretation to understand the income equality pattern of the farmers. Further, decomposition analysis of Gini coefficient has been employed to quantify the contribution of various income sources to the total inequality and the signs thereof. This decomposition analysis is useful to guide the prioritisation of various sources of income component to derive equitable distribution. To examine the contribution of different sources in income inequality, source-wise decomposition of Gini index was estimated (Lerman and Yitzhaki, 1985). Denoting total household income by y, the cumulative distribution function for total household income by F(y), which takes a value of 0 for the poorest household and 1 for the richest, and the mean total household income across all households by $\overline{y}$, the Gini index can be decomposed as follows:

$$G_y = 2Cov[y, F(y)]/\overline{y} = \sum S_i R_i G_i \qquad (1)$$

Where, G_y is the Gini index for total income, G_i is the Gini index for income y_i from source i, S_i is the share of total income obtained from source i, and R_i is the Gini correlation between income from source i and total income. The Gini correlation is defined as:

$$R_i = Cov\,[yi, F\,(y)] \,/\, Cov\,[yi, F\,(y_i)] \qquad (2)$$

Where, $F\,(y_i)$ is the cumulative distribution function of household income from i^{th} source. The Gini correlation R_i can take values between -1 to +1. The overall (absolute) contribution of source of income i to the inequality in total household income is thus $S_iR_iG_i$. When the income source is a constant, then R will equal 0, implying that the source's share of the Gini is 0. As such components raise their share of total income, overall inequality falls.

A key rationale for studying decompositions by source is to learn how changes in particular income source will affect overall income inequality. Consider a change in each household's income from source i equal to ey_i, where e is close to 1. Starting from equation (1), we can derive a clear expression for the partial derivatives of the overall Gini with respect to a percentage change e in source i. The derivation yields:

$$\frac{\delta G_y}{\delta e_i} = S_i\,(R_iG_i - G_y) \qquad (3)$$

Dividing equation (3) by G_y yields the source's marginal effect relative to the overall Gini, which can be written as the sources, inequality contribution as a percentage of the overall Gini minus the source's share of total income:

$$\frac{\delta G_y/\delta e_i}{G_y} = \frac{S_iG_iR_i}{G_y} - S_i \qquad (4)$$

3 Agricultural Development Status

Land Utilisation and Distribution of Farm Lands

The northeastern states are severely constrained by difficult terrain with 70 per cent area under hills, and forest covering ranges between 42 and 76 per cent of the area of each states (Mohan, 2003). Agricultural performance of the region remains sluggish on account of a number of factors. For one, the ratio of irrigated area to total area under all crops in the NER has generally remained much lower than the all-India figure of 33 per cent. Moreover, the total area in the NER covered under high-yielding varieties (HYVs) formed only around 0.25 per cent of the all-India total in 1996-97 (Table 3.1).

Assam is having nearly 43 per cent of total geographical area suitable for agricultural operation, followed by Nagaland (38%), Tripura (30%), Mizoram (21%), Sikkim (16%) and Manipur (7%). Out of the total arable land (6.4 million ha) a little over 60 per cent (4 million ha) area has been brought under net sown area (NSA) at the northeast region level, while at national level nearly three-fourth (72%) area has been brought under actual cultivation (NSA). It is due to inherent characteristics of terrain, fragility, inaccessibility and traditional land tenure system. There was moderate increase in net sown area in most of the northeastern states during 1975-76 to 1995-96. This shows that land was either made suitable for cultivation through land development activities or was considered suitable (exploitation motive) for cultivation in these states. The area under forest in northeastern states accounted for major share (66% to total geographical area of NER) in its land use pattern. At all-India level the share of forest area has been estimated to be 23.4 per cent (78.37 million ha) of the total geographical area (Government of India, 2009). Area (in sq km) in northeastern states under forest cover is, Arunachal Pradesh (67,472),

Table 3.1

Land Utilisation Pattern in Northeastern States of India (2000-01)

State	*Total Geographical Area ('000 ha)*	*Arable Land ('000 ha)*	*Net Sown Area ('000 ha)*	*Gross Cropped Area ('000 ha)*	*Cropping Intensity (per cent)*	*Gross Irrigated Area ('000 ha)*	*Fertiliser Use (Kg/ha, 2003-04)*	*Average Holding Size (ha)*
(1)	(2)	(3)	(4)	(5)	(6)	(7)	(8)	(9)
Arunachal Pradesh	8374.3	293(4)	164(56)	263	160	36(14)	2.8	3.31
Assam	7843.8	3387(43)	2734(81)	4065	149	572(14)	46.6	1.17
Manipur	2232.7	164(7)	140(85)	209	149	86(41)	130.5	1.22
Meghalaya	2242.9	1074(48)	230(21)	277	120	105(37)	17.0	1.33
Mizoram	2108.1	445(21)	94(21)	94	100	13(14)	-	1.29
Nagaland	1657.9	626(38)	300(48)	314	105	65(21)	2.2	4.82
Tripura	1048.6	310(30)	280(90)	428	153	24(6)	29.4	0.60
Sikkim	710	110(16)	95(86)	126	133	16(13)	3.5	1.65
Northeast region	26218.3	6409(24)	4037(63)	5776	143	917(15)	-	-
All India	328726.3	194680(59)	141101(72)	187009	133	61659(33)	89.8	1.41

Note: Figures in parentheses in column (3), (4) and (7) indicate the percentage of arable land to total geographical area, percentage of net sown area to total arable area and percentage of irrigated area to gross cropped area, respectively for the northeastern states.

Source: CMIE (1997 and 1998), *Basic Statistics of NER 2006* and *Economic Survey* (various issues).

Assam (27,758), Manipur (16,952), Meghalaya (17,205), Mizoram (18,600), Nagaland (13,665), Sikkim (3,357), Tripura (8,173) and NER (1,73,182). All-India forest area is 6,90,171 thousand ha. It is because of the inherent characteristic of favourable agriculture and topographical condition in northeastern states. There is a sharp increase in area under forest in Nagaland. The reverse was true in the case of area not available for cultivation. It could be further, seen from Figures (3.1-3.2) that the cropping pattern in NE states remained highly specialised in food grains.

Among the northeastern states, Tripura ranked first in terms of percentage of NSA to total available arable land (90%), followed by Sikkim (86%), Manipur (85%), Assam (81%), Arunachal Pradesh (56%), Nagaland (48%), Mizoram (21%) and Meghalaya (21%). The gross cropped area (GCA) of the region is over five million ha, out of which Assam alone is accounting for more than 70 per cent of the area. The cropping intensity of the region is marginally high (143%) as compared to the nation as a whole (133%). Overall, the development of irrigation facility is very poor and nearly 86 per cent of the total cropped area is being cultivated under rainfed condition. The percentage of gross irrigated area (GIA) to GCA is maximum in Manipur (41%), but with a low magnitude followed by Meghalaya (37%), Nagaland (21%), Mizoram (14%), Arunachal Pradesh (14%), Assam (14%), Tripura (6%) and Sikkim (13%), which indicates the need of creation of massive water harvesting and storing structures suitable to the physiographic situation of the region. The fertiliser consumption in NER region showed that the consumption per ha is hovering only around 35 kg (2003-04). Manipur is the exception among northeastern states in terms of using fertiliser at higher level (130 kg/ha). Next to Manipur, Assam is using maximum amount of fertiliser per ha (46.6 kg/ha) followed by Tripura (29.4 kg/ha), Meghalaya (17 kg/ha), Sikkim (3.5 kg/ha), Arunachal Pradesh (2.8 kg/ha), and lowest in Nagaland (2.2 kg/ha) observed during 2001-02. Also, the trend in use of fertiliser during recent period (2001-02 to 2003-04) does not show any significant difference within the northeastern states. The average size of farm holdings among northeastern states was observed to be highest in Nagaland (4.82 ha), followed by Arunachal Pradesh (3.31 ha), Sikkim (1.65 ha), Meghalaya (1.33 ha), Mizoram (1.29 ha), Manipur (1.22 ha), Assam (1.17 ha) and lowest in Tripura (0.60 ha). At regional level, three out of eight states from this region, namely, Arunachal Pradesh, Nagaland and Sikkim have the higher farm holding size as compared to the national average (1.41 ha). However, in spite of possessing larger farm holdings, the hill farmers can not enjoy the economies of scale due to limited substitutability among the factors of production (e.g. replacing human labour with machine labour), hence, the scope for reduction in cost of production is also restricted.

The land utilisation pattern for agriculture in NER region is clearly indicating the existing wide gap between the amount of land available for agriculture (arable land) and the land actually brought under cultivation (NSA). Mostly the lands are used for a single season and only 37 per cent of NSA is being utilised for cultivation of more than once. Despite of receiving highest rainfall, the region is predominated by rainfed agriculture; the agricultural operation is severely restricted by either heavy rainfall or lack of soil moisture. The pace of agricultural development in NER, in terms of land intensification, is quite slow, perhaps the major factor that the region is still carrying rich biodiversity. The development of agriculture plan should be given top priority to protect the natural wealth, therefore, needs to be addressed from the purview of multifunctional (e.g., contribution to ecology, maintaining rich biodiversity etc.) nature of agriculture.

Trends in Food Grains Production and Productivity

The trend in food grains production for the northeast region as a whole observed to be declining over preceding decades (Table 3.2). During 1980-81 to 1989-90 period the food grains production growth was estimated to be 1.52 per cent whereas the same declined at 1.25 per cent during 1990-91 to 1999-2000 period. Similar to food grains production, rice production was also observed to be declining during the same period (from 1.41 per cent and 1.24 per cent during 1980-81 to 1989-90 and 1990-91 to 1999-2000 period, respectively). Overall, the food grain production for the region has increased by 32 per cent during 1980-81 to 1989-90 (from 4,138 thousand tonnes to 5,449 thousand tonnes) whereas the same was only 16 per cent during the later decade i.e., 1990-91 to 1999-2000 (from 5,449 thousand tonnes to 6,309 thousand tonnes). Assam is the single largest producer within the northeastern states contributing more than 60 per cent of the total food grains production. Positive trend has been observed for the states of Manipur and Nagaland.

In other northeastern states the food grain production showed both positive and negative kind of trend. Concerted efforts are needed to increase the food grain production in the region to make it food-wise self-reliant through promotion of suitable HYVs along with the complete package of practices.

Table 3.2

Trend in Food Grains Production in the Northeastern States of India

States	*Annual Compound Growth Rate (%) during*			*Average Production ('000 tonnes) during*		
	1980/81 to 1989/90	*1990/91 to 1999/2000*	*2000/01 to 2005/06*	*1980/81 to 1989/90*	*1990/91 to 1999/2000*	*2000/01 to 2005/06*
Arunachal Pradesh						
Rice	5.42	-1.04	1.46	117	128	143
Food grains	5.19	-0.83	2.01	172	208	231
Assam						
Rice	1.08	1.06	-2.45	2548	3365	3749
Food grains	0.96	1.07	-2.56	2737	3544	3902
Manipur						
Rice	0.33	2.77	1.65	270	351	384
Food grains	0.34	2.79	1.44	284	362	396
Meghalaya						
Rice	-2.20	4.08	-1.97	120	131	184
Food grains	-1.18	3.59	-2.23	151	165	2218
Mizoram						
Rice	8.54	4.80	-0.62	42	94	106
Food grains	7.91	4.12	0.76	49	115	129
Nagaland						
Rice	3.00	0.48	3.02	104	170	244
Food grains	3.31	1.88	5.31	122	225	384
Sikkim						
Rice	7.65	-0.23	-0.21	16	22	22
Food grains	8.52	-1.81	0.31	90	105	100
Tripura						
Rice	2.16	1.06	-0.02	401	486	553
Food grains	2.11	0.87	0.04	410	499	564
NE region						
Rice	1.41	1.24	-1.31	3722	4918	5629
Food grains	1.52	1.25	-0.88	4138	5449	6309

Source: Authors' calculation based on *Basic Statistics of North Eastern Region* (Various issues).

The magnitude of change in trends in food grains productivity was higher during 1980-81 to 1990-91 periods as compared to 1990-91 to 2002-03 periods in various northeastern states, except for Meghalaya (Table 3.3). The improvement in food grains productivity was highest in Nagaland (36%), during 1980-81 to 1990-91 followed by Tripura (32%), Mizoram (24%), Manipur (22%), Assam (19%) and Arunachal Pradesh (11%). During later decade (1990-91 -2002-03), Mizoram recorded highest increment in productivity (44%) followed by Meghalaya (42%), Nagaland (35%), Manipur (31%), Tripura (23%), Assam (12%) and Arunachal Pradesh (5%). It is notable from Table 3.3 that the trend in food grains productivity was almost similar to the trend of change in rice productivity for all northeastern states. The reason being, rice alone accounted for more than 70 per cent of the cropped area in these states and also has almost equal share to the total food grains production. Therefore, productivity of rice observed to be almost determining the productivity of food grains as a whole. Given the positive improvement of food grains production, the major challenge is to continue this improvement, particularly in view of the complex, diverse and risk-prone (CDR) production environment existing in these hilly states. Rice in these states are grown traditionally and characterised by low yield, particularly at upland situation. Moreover, some of the northeastern states possess huge rice diversity where advancement of modern varieties is almost insignificant. The rice in this region is cultivated under varied topographical situation and production environment, for example, Assam has been dominated by rainfed production system, which covered 79 per cent of total cropped area. The rainfed area were further classified into upland and shallow land, which occupied 60 per cent of total rice area, deep water 14 per cent and areas with intermittent water (30-100 cm) was 26 per cent. Rice in Assam has a dubious distinction, having grown to diverse land types in three seasons in a year. This diversity of production ecosystems made the understanding of rice production system more complex (Bhowmick *et al.*, 2005). The northeast region, the area under modern rice variety is less and a large number of traditional varieties are being grown by the farmers. The traditional farming systems are being followed and most of the farming practices are operating in subsistence to semi-commercial stage. Additionally, the farming practices are not responsive to the market

Table 3.3

Trends and Progress in Food Grains Productivity in Northeast Region of India (1980-81 to 2002-03)

(q/ha)

States/Year	*1980-81*	*1990-91*	*2002-03*	*Change (%)*	
				1980-81/ 1990-91	*1990-91/ 2002-03*
(1)	*(2)*	*(3)*	*(4)*	*(5)*	*(6)*
Arunachal Pradesh					
Rice	10.6	11.7	12.2	10.38	4.27
Food grains	10.5	11.7	12.3	11.43	5.13
Assam					
Rice	11.0	13.1	14.7	19.09	12.21
Food grains	10.7	12.7	14.2	18.69	11.81
Manipur					
Rice	14.5	17.4	23.8	20.00	36.78
Food grains	14.5	17.7	23.1	22.07	30.51
Meghalaya					
Rice	13.4	11.6	17.2	-13.43	48.28
Food grains	14.3	11.5	16.3	-19.58	41.74
Mizoram					
Rice	9.8	12.4	19.1	26.53	54.03
Food grains	1045	13.0	18.7	23.81	43.85
Nagaland					
Rice	9.0	12.3	14.9	36.67	21.14
Food grains	851	11.6	15.7	36.47	35.34
Tripura					
Rice	13.6	18.3	22.7	34.56	24.04
Food grains	13.5	17.8	21.9	31.85	23.03
Sikkim					
Rice			14.4	-	-
Food grains			13.2	-	-

Source: Compiled from *Basic Statistics of North Eastern Region – 2002* and various issues, CSO Database - *Statistical Abstract* (2003).

demand, as most of the landraces are traded through small-scale informal marketing channels. The primary objective of growing rice is not for profit maximisation but to meet the household requirements to maintain households' food security (Datta *et al.*, 2007). The opportunity of *in situ* on farm conservation may be quite high and feasible in this region because the farmers are continuing to grow a large number of landraces since ages. So, conservation of the traditional varieties can be viewed as a unique opportunity for this region, which has great importance for our society (public value) to secure the future varietals development programme. The efforts to conserve plant genetic resources have long been emphasised and northeastern states particularly, Arunachal Pradesh, Nagaland, Mizoram and Meghalaya have the right potential for exploring these opportunities.

Food Security

To achieve the food security, the growth rate for food grains production should exceed to sustain the population growth rate at individual state level. But this is not happening for the states, Arunachal Pradesh, Assam, Nagaland, Tripura and Sikkim (Table 3.4). However, rest of the states in this region, Manipur, Meghalaya and Mizoram have been able to maintain higher food grains production growth as compared to population growth rate. Thus, relatively food availability in these states are improving. The performance of food grains production growth was experiencing to be alarming for Sikkim (-1.81%), which needs to be looked into seriously. Similarly, also the per capita output from agriculture (excluding livestock and fisheries) was observed to decline for Sikkim (from Rs. 3,549 in 1991 to Rs. 2,372 in 2001) and Arunachal Pradesh (from Rs. 5,062 in 1991 to Rs. 2,333 in 2001). It is important to identify reasons in each decline. The lack of data prevented us doing such analysis. As far as per capita food production is concerned, Arunachal Pradesh, Sikkim and Tripura failed to maintain the food production level during 2001 as compared to 1991. As far as per capita food production is concerned, Arunachal Pradesh, Sikkim and Tripura failed to maintain the food production level during 2001 as compared to 1991. High population growth rate is the major factor for this decline, however, the magnitude of decline was very high for Arunachal Pradesh (from 679 g to 501 g per capita production) and Sikkim (from 718 g to 539 g) that needs to be looked into seriously.

Table 3.4

Growth Rate of Population, Food Grains Production, Per Capita Income vis-à-vis Change in Per Capita Food Grains Availability and Per Capita Agricultural Output

Particulars	*Population Growth (%)*	*Food Grains Production Growth (%)*	*Per Capita Availability of Food Grains (gram/day)*		*Per Capita Income Growth (%) (at 1993-94 Constant Prices)*	*Per Capita Output (Rs.) from Agriculture (at 1993-94 Constant Prices)*	
	1991-2001	*1990-91 to 1999-2000*	*1991*	*2001*	*(1993-2003)*	*1991*	*2001*
Arunachal Pradesh	2.33	-0.83	679	501	0.94	5062	2333
Assam	1.73	1.07	421	429	1.09	2087	2260
Manipur	2.63	2.79	425	434	4.38	1457	1641
Meghalaya	2.62	3.59	236	241	4.84	1455	1575
Mizoram	2.56	4.12	304	381	-	1398	2151
Nagaland	4.97	1.88	447	382	2.80	1669	3210
Tripura	1.46	0.87	511	449	8.02	1949	2616
Sikkim	2.72	-1.81	718	539	3.51	3549	2372

Note: * Data pertains to the period of 1989-90 to 2002-03.

Source: Estimated based on data from *Basic Statistics for North Eastern Region 2002*, CSO Database *on State wise Estimate of Value of Output from Agriculture and Livestock*.

The achievement of higher growth rate in food grains production over and above population growth rate is important for attaining food security, but, simultaneously sustaining the per capita income growth is also a must to ensure the food availability at household level. The growth of per capita income reflects the purchasing power of the individuals which is indispensable to ensure the distribution of available food grains. Therefore, food grains production growth ensures the availability of food in one hand and per capita income growth guarantees the accessibility of food produced on the other. The results in Table 3.4 indicated that growth rate in per capita income (during 1993-94 to 2002-03 at 1993-94 constant prices) was highest in Tripura (8%) followed by Meghalaya (5%), Manipur (4%), Sikkim (3.5%), Nagaland (3%), Assam (1%) and Arunachal Pradesh (0.9%). Results further indicated that some of the northeastern states (Manipur, Meghalaya and Mizoram) showed positive improvement. Therefore, the problem of food security for these states can be attributed partly to the deficiency in food production and also due to the distribution failure. Besides improving the food grains productivity, efficient public distribution system is equally essential to achieve the desired food security for this region.

Cropping Pattern

The cropping pattern in the northeastern states is predominated by the food crops, which accounted for around 68 per cent of total cropped area (Table 3.5). Out of which rice alone is accounted for 59 per cent. More than 80 per cent of total cropped area is under food grains for the states of Manipur and Nagaland. Other northeastern states accounting for around 60-80 per cent of the cropped area under food grains crop except for Meghalaya (51%). The cropping pattern in the northeastern states as a whole is gradually changing in favour of non-food crops such as fruits and vegetables but the magnitude of change is almost insignificant. For example, during 1991-92 around 71 per cent cropped area was under food grains and 14 per cent area under horticultural crops. Presently the area under food grains crop has come down at 68 per cent but the area under horticultural crops remains almost unchanged. Marginal change in area under food grains has been shifted to other crops. Individually these crops accounted for very meagre area. The region accounted for nearly 4 per cent of country's horticultural total area. Area under fruits, vegetables and spices estimated to be around 294 thousand ha, 328 thousand ha and 162 thousand ha, respectively, which are accounting for 6 per cent, 5 per cent and 3 per cent of total country's area under the respective crops (Table 3.6). Cropping pattern in the northeastern states clearly indicates the importance of high value horticultural crops particularly in the states of Sikkim, Manipur, Arunachal Pradesh, Meghalaya, Mizoram and Nagaland. Horticultural sector has adequate growth potential to increase the overall agricultural output of the region. Diverse agroeconomic conditions of these states are highly suitable for growing high value horticultural crops such as, cardamom, ginger, turmeric, pineapple, orange, passion fruits, flowers etc., which have large export potential also.

Table 3.5

Area under Food Grains Crops in the Northeastern States of India (2002-03)

States	*Area under ('000 ha)*			*Gross Cropped Area (GCA)*
	Rice	*Food Grains*	*Oilseeds*	
Arunachal Pradesh	125 (47)	196 (74)	24	264
Assam	2541 (62)	2749 (67)	297	4093
Manipur	159 (80)	170 (85)	2	199
Mizoram	57 (63)	69 (76)	7	91
Meghalaya	106 (40)	135 (51)	9	266
Nagaland	151 (51)	248 (84)	60	295
Sikkim	-	77 (64)	10	121
Tripura	240 (57)	254 (60)	6	420
Northeast region	3379 (59)	3898 (68)	415	5749
All India	40279 (22)	111494 (60)	21224	187009

Note: Figures in parentheses indicates per cent to total gross cropped area (GCA).

Source: Compiled based on database from *Basic Statistics of NER 2006*.

Table 3.6

Area under Horticultural Crops in the Northeastern States of India (2004-05)

States	*Area under ('000 ha)*			
	Fruits	*Vegetables*	*Spices*	*Horticulture (Total)*
Arunachal Pradesh	50.7	20.4	6.61	78
Assam	93.3	194.5	71.08	359
Manipur	51.2	13.4	17.33	82
Mizoram	21.2	5.7	6.29	33
Meghalaya	23.8	32.7	18.88	75
Nagaland	8.2	17	3.55	29
Sikkim	13.3	11.9	33.77	59
Tripura	32.2	32.7	4.46	69
Northeast region	294 (6)	328 (5)	161.97 (3)	784 (4)
All India	4963.8	6755.6	5908.7	17628

Note: Compiled based on database from *Basic Statistics of NER 2006*, National Horticulture Board (2006) and Spice Board (2007).

Figure 3.1

Cropping Pattern in Northeastern Region during 1991-92

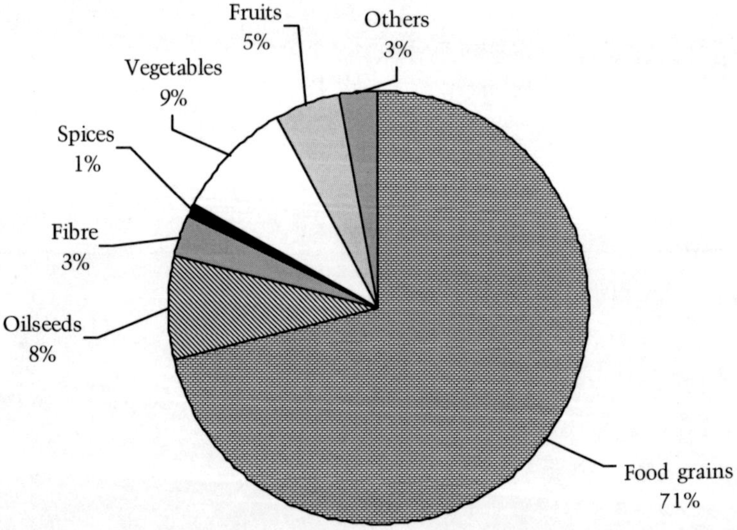

Source: Authors' compilation (2008).

Figure 3.2

Cropping Pattern in Northeastern Region during 2000-01

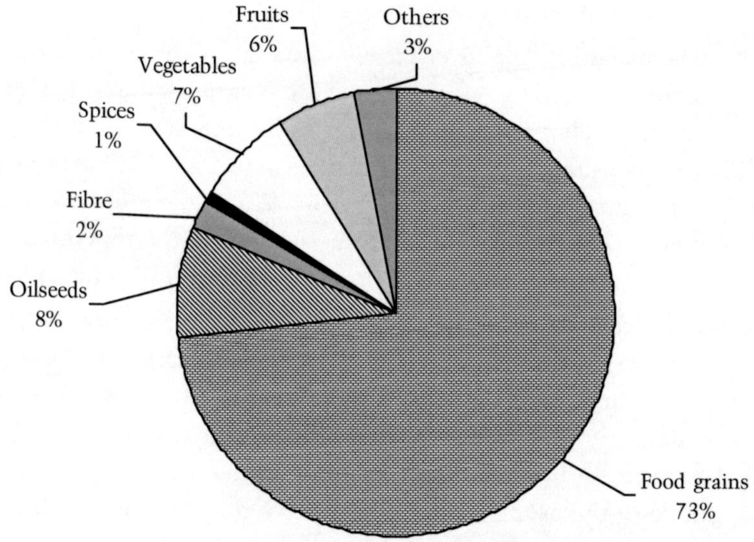

Source: Authors' compilation (2008).

Scope of Crop Diversification

Northeastern states are predominated by cereal crops and nearly 70 per cent of the agricultural land is covered by food grains only (Figure 3.1 and 3.2). It is assumed that diversification towards other high value crops is likely to improve the farm income. Under the present analysis the diversification at state level has been estimated through Simpson index. Extent of crop diversification has been estimated by employing Simpson Diversification Index (SID) as $SID = 1 - \Sigma (X_i / \Sigma X_i)^2$. Here, i = area under i^{th} crop. The index value ranges in between 0 to 1, as the estimated value approaches from 0 to 1. The salient observations about crop diversifications are:

1. At northeastern region level the extent of crop diversification is very low (0.37 only) with 79 per cent are under rice only. Low crop diversification was estimated to be in the state of Tripura and Manipur.

2. Extent of diversification was relatively more in Arunachal Pradesh (0.59-0.62) and Meghalaya (0.57-0.58) but the agricultural performance in these states are considered to be poor. Moderate crop diversification was observed under the states Nagaland (0.44-0.57), Mizoram (0.42-0.46) and Assam (0.34-0.33).
3. The extent of crop diversification was not observed to be positively correlated with the value of agricultural output—implying that more critical analysis is required to prescribe crop diversification policies for the region to improve the farm income. Rather it can be assumed that suitable farming system might be a viable proposition for the region. Diversification analysis also indicates that higher crop diversification is not always ensuring a higher value of crop output.

The possible reason might be that some of the crops with low share in total crop area are fetching better prices and increasing the value of agricultural output substantially without altering the share in total crop area significantly. Therefore the value of agricultural output is increasing but the extent of diversification remaining unchanged or insignificant. At aggregate level the extent of crop diversification is quite low in the north-eastern states but at micro-level the studies of farming systems indicates very high level of crop diversification particularly at highland situations. This is a typical dichotomy prevailing in the northeast agriculture. The reason being, under upland situation a variety of crop mix are cultivated consisting of horticultural crops such as fruits, vegetable and spices. These crops are considered to be high value crops and serve the purpose of cash needs of the farmers. Mostly these are cultivated at a very small scale and the economies of scale are also restricted due to the fact that these farm lands are primarily managed by the family labour and the scope of farm mechanisation is limited. Moreover very high level of crop diversification likely to lead lowering of crop income due to more risk association with various activities as well as management factor. Also supply of quality seeds and other essential inputs and delivery of output at remunerative prices would be a pivotal factor to limit the crop diversification particularly under the geographically difficult terrains of the region. Therefore, careful planning is required to lift the farm income through improvement of present farming systems practiced by the farmers and gradually pulling the farmers to semi-commercial stage from the subsistance farming followed by the commercial farming. Otherwise suitable site or locations can be identified for production of specified agroproducts to increase the marketable surplus so as to facilitate the agricultural marketing through private participation.

Value of Agricultural Output and Growth

The value of output growth for the fruits and vegetables was observed to be substantially higher in all northeastern states (except Arunachal Pradesh and Assam) in comparison to all-India level (5.43%). Nagaland has recorded more than 20 per cent annual compound growth rate for total fruits and vegetables production followed by Mizoram (15%), Tripura (10%), Meghalaya (7%), Sikkim (6%) and Manipur (5%) during the period 1990-91 to 2002-03. Within the agriculture sector horticulture sector has been showing substantial growth potential across the northeastern states. The investment in horticultural sector in the northeastern states has started paying dividend but at a slower rate in comparison with its underlying potential. Development of fruit sector has been observed to be encouraging but same is not true for the growth in vegetable production. And the spices production have been growing at a slow rate or remaining almost unchanged across the various states of the region. Mass production and distribution of quality planting material for fruits and spices can be given top priority. Special support and focus needs to be given to capacity building of the farmers to produce vegetables under protected as well as open condition. Expansion of area under off-season vegetable crops should be made around the various hilly tracts of the region. Appropriate methodologies are needed to estimate the area, production and productivity of horticultural data and simultaneously, creation of detailed database is strongly advocated. Finally, establishment of market intelligence unit is utmost important to provide advisory services of price movement and market arrival of various horticultural commodities through recording and analysing the day-to-day, weekly, monthly, seasonally or yearly data.

Annual compound growth rate of value of agricultural output observed to be quite encouraging for the period from 1990-91 to 2002-03 (Table 3.7). The growth performance was even better than national level (2%) in all northeastern states except, Arunachal Pradesh (-4.3%) and Sikkim (0.31%).

Table 3.7

Annual Compound Growth Rate of Value of Agricultural Output in Northeastern States of India (1990-91 to 2002-03)

(per cent)

State/Commodity	*Rice*	*Pulses (Total)*	*Cereals (Total)*	*Fruits & Vegetables*	*Spices & Condiments*	*Agriculture (Total)*
(1)	*(2)*	*(3)*	*(4)*	*(5)*	*(6)*	*(7)*
Arunachal Pradesh	0.39	3.25	0.42	-11.13	13.51	-4.32
Assam	1.88	-0.62	1.82	3.46	0.87	2.63
Manipur	1.62	6.63	1.61	4.16	5.45	2.46
Meghalaya	5.56	2.04	5.12	6.67	3.32	2.22
Mizoram	3.40	-3.77	3.30	15.26	14.55	4.72
Nagaland	2.99	6.68	4.67	20.33	23.58	12.57
Tripura	1.86	-1.49	1.79	21.02	8.53	5.22
Sikkim	0.69	-4.66	-0.12	5.55	-3.97	0.31
India	1.04	-0.35	1.26	5.43	4.36	2.02

Source: Estimated from CSO Database on *State-wise Estimate of Value of Output from Agriculture and Livestock* (at 1993-94 constant prices).

The growth rate observed to be highest for Nagaland (12.57%) followed by Tripura (5.22%), Mizoram (4.72%) and positive for all other states. The growth rate of Nagaland can be attributed to the high growth rate achieved for production of spices and condiments (24%) and fruits and vegetables (20%). Growth rate for value of rice output continues to be positive for all the states at varying magnitude, but output growth for value of pulse production shows some dismal figures, except for Nagaland, Manipur, Arunachal Pradesh and Meghalaya, where the growth rates were positive. Pulse production can be increased in the state with adequate technological back-up and extending complete package of practices in a mass scale. Similar to output growth for rice, total cereal production also indicated positive growth for the northeastern states, except Sikkim. The growth rate for value of output from fruits and vegetables as well as spices and condiments were observed to be highly encouraging (except for Arunachal Pradesh and Sikkim, respectively). This calls for highly focussed policy initiation to realise the full potentialities. Major task would be to increase the marketable surplus through horizontal integration and then, promoting vertical integration for value addition of these products. To facilitate this integration, there is a need for credible database at micro-level on these products giving details of area-wise volume of production and their movement, cost of production and possible value addition. Once this integration is successful, thereafter, the possibility of air lifting for these value-added products can be thought of to reduce the time for transportation.

Value of Output from Livestock and Growth

Similar to agriculture, the output growth for livestock product also observed to be encouraging for the northeastern states (Table 3.8). The growth rate was varying from highest in Nagaland (6.63%), followed by Tripura (6.37%), Meghalaya (5.93%), Mizoram (5.61%), Manipur (2.09%), Sikkim (1.83%), Arunachal Pradesh (1.54%) and Assam (0.70%). Four out of eight northeastern states have even recorded much higher output growth as compared to all-India level (3.79%). The major sources of growth were realised from meat and milk production during this period (1990-91 to 2002-03). Value of output growth from egg showed both positive as well as

Table 3.8

Annual Compound Growth Rate of Value of Output from Livestock and Livestock Products in Northeastern States of India (1990-91 to 2002-03)

(per cent)

State/Commodity	*Milk*	*Meat*	*Egg*	*Dung*	*Livestock (Total)*
(1)	(2)	(3)	(4)	(5)	(6)
Arunachal Pradesh	1.17	0.50	-6.31	5.72	1.54
Assam	0.81	1.72	0.65	-0.47	0.70
Manipur	-1.84	4.59	-1.19	-6.27	2.09
Meghalaya	2.85	7.69	2.02	1.85	5.93
Mizoram	6.84	5.34	5.31	3.39	5.61
Nagaland	2.69	8.97	3.78	3.08	6.63
Tripura	8.77	9.64	11.53	2.47	6.37
Sikkim	3.09	-3.84	-3.00	-1.72	1.83
India	4.14	3.44	13.32	0.62	3.79

Source: Estimated from *CSO* Database on *State wise Estimate of Value of Output from Agriculture and Livestock* (at 1993-94 constant prices).

negative sign. The northeastern states are quite suitable for rearing diverse livestock population due to abundance of free grazing land. Livestock such as cattle, buffaloes, goat, sheep, poultry, pig and in addition *mithun* and yak are reared by the farm families in a traditional way. The livestock sector needs to be made more vibrant through enhancing the productivity, particularly, in pockets where the livestock densities are very high with respect to existing population base, available workers or net sown areas. It is essential to increase the volume of marketable surplus for trading of livestock products within the countries as well as to neighbouring countries, where the demand for livestock products are always exceeding supply. However, to achieve the surplus production suitable technological imperatives as well as horizontal integration among the primary producers is a must.

Employment and Output from Agriculture

Agriculture and allied activities is the major employment provider for the workers in this region, which is accounted for nearly 50-70 per cent of the total workforce across the various states (Table 3.9). But the contribution of agricultural output is quite low, ranging in between 27-38 per cent only, which is almost half of the share of labour force, engaged in this sector, implying the low level of labour productivity. The efficiency of labour input must be increased through accelerated agricultural development, for which identification and prioritisation of the potential sub-sectors is extremely important. Also the linkages between agricultural sector and other sectors needs to be analysed to employ the surplus labour gainfully. The agricultural output can be increased through promotion of possible horizontal and vertical integration of the agriculture and allied activities. It is important to note that the female workers are mostly engaged in agricultural activities and in fact, nearly 70-80 per cent of the female workers are earning their wages from this sector, whereas, nearly 50-60 per cent of the male workers are engaged in agricultural activities. Women provide an important source of labour for cash crop production, growing food for household consumption, raising livestock and generating additional income through diversified activities. However, much of this work remain statistically 'invisible'. Nevertheless, the women group needs to be given special attention to increase their efficiency, which in turn will lead to improvement in the overall labour productivity and will increase the share of agricultural output to net state domestic product (NSDP). When credit and resources are unavailable, reducing the variability of household earnings generally entails choosing less efficient method of production and thus realising low income. This trade-off occurs most oftenly in female headed household where resource constraints are greatest, thus as a consequence of their restricted range of choices, women tend to retain traditional modes of economic activities (Todaro and Smith, 2004).

Value of agricultural output per worker (cultivators plus agricultural labourers) indicated some encouraging results and in fact, values for some of the states exceed the respective per capita income when compared at current prices of 2000-01. For example, per capita income (state level) of all northeastern states at current prices of 2000-01, was below their respective value of agricultural output per worker. This implied that even with the low level of labour productivity, the agricultural activities continued to provide relatively better employment opportunity as compared to overall activities available in these states. However, this may have another implication also, that is the opportunities of gainful employment in non-agricultural activities are restricted or not available. Therefore, the results indicated that agricultural activities continue to provide employment opportunities to majority of the workers and might be given priority for overall economic development of these states.

Table 3.9

Employment and Output from Agriculture Sector in Northeast Region (2001)

Particulars	*Employment in Agriculture (% to Total Workers)*			*Share of Agricultural Output to NSDP (%)*	*Value of Agricultural Output per Worker* (Rs./Year)*		*Per Capita Income at Current Prices*
	Male	*Female*	*Total*		*Current Prices (2000-01)*	*Constant Prices (1993-94)*	
(1)	*(2)*	*(3)*	*(4)*	*(5)*	*(6)*	*(7)*	*(8)*
Arunachal Pradesh	50	81	62	37	21395	12356	15538
Assam	51	57	53	38	24669	13213	12797
Manipur	56	60	57	35	14347	11667	12369
Meghalaya	61	73	66	27	17347	8478	16100
Mizoram	54	68	60	29	16847	—	17826
Nagaland	59	80	68	31	18381	12995	—
Tripura	46	63	51	28	23924	8655	15983
Sikkim	47	71	56	28	14211	12702	16077

Notes: *Includes cultivators and agricultural labourers.

Source: Calculated from Census of India (2001), RBI (2005). *Handbook of Statistics on Indian Economy* and CSO Database—*Net State Domestic Product (1993-94 Series)*.

4 Rice in the Northeast Region

Yield Gap Analysis and Prospects

Rice and maize are the major cereal crops in this region. All the farmers prefer to grow rice, as it is their staple food crop. Since rice production in NER in general is consumption oriented to cater to the needs of the farmers and local population and not for marketing. Under the traditional procedures for growing rice, the choice of alternatives open to a farmer is extremely limited. Since economics involves the comparison of choice among alternatives, there are few economic decisions to be made in this subsistence environment. Though rice is the major crop and occupies about 59 per cent of the total cropped area but the region has the stigma of low and uncertain productivity in rice and slow and poor dissemination of new technologies. In such environment, the yields are not only low but also fluctuate abnormally. In other words, there is no significant improvement in the productivity of food grains in such adverse environments. Although the region is lagging far behind than irrigated and other favourable regions, this region has a considerable potentiality and opportunities, as it possesses fairly good soil, humus, high precipitation, enough human resource, large cattle population, etc. In general the common opinion is that the introduction of modern package of practices along with high yielding varieties (HYVs) will help to increase productivity particularly in rice. But the introduction of exotic HYV rice in NER shows that instead of increasing productivity of rice, these varieties demanded higher inputs. The resultant outcome is growth of intensive foliage and the higher nutrient contents of the plants become a good host for insect pests and disease pathogens. Along with this problem, congenial climatic conditions like high rainfall, higher humidity throughout the year doubled the problem of pest infestation. They are not able to capitalise on the natural predators that occur at a given time in the season. It has been observed that cultivation of some of the blast resistant varieties under poor nutrient management conditions helps to increase the incidence of brown leaf spot. Concerns are also raised on increased evidences as well as emergence of new insect pests and disease pathogens that are generally reported from northeastern hill region and the ultimate consequence is reflected on the productivity of rice. Rice in the region confronts many adversities and diversities, posing a challenge to agricultural policymakers and researchers. Evidences suggest that gap in research challenges has been bugling out, which also created more demand for rice research in the region (Jha and Pandey, 2005).

A perusal of the literature showed that there are lot of works that have been done both in irrigated (Roy and Datta, 1999) and rainfed (Ramaswamy *et al.*, 1996; Widawsky and O'Toole, 1996) ecosystem. Those studies have seldom identified in homogeneous production zones where the constraints were of more or less similar in nature and research outputs have homogeneous impacts. But the hilly region of northeastern region possess varying degree of topographical situation and simultaneously region constrained with low productivity because the agriculture is governed by complex, diverse and risk-prone (CDR) nature and mostly cultivated under rainfed condition. Moreover due to inaccessibility (a product of altitude and terrain), fragility (a product of steep slope and other biophysical conditions) and marginality (a product of nature and manmade factors having both physical and socioeconomic dimension) of the region, it is very difficult to transmit the technology in a broader scale. As an example it is assumed that the variety contributes about 40 per cent towards increased productivity, but the region failed to make any impact due to wide diversity in rice ecology and very low solar radiation throughout the growth process of rice production. In the mid altitude cold is the major constraint during flowering stage, similarly in the high altitude, cold is the major constraint both at vegetative as well as flowering stage. The magnitude of climatic risks has been so acute that the farmers not only lose the returns but also lose the production cost/investment. Productivity of rice in the region is directly linked with different grades of land. For example in the high altitude upland area, productivity is least whereas productivity is modest in mid altitude terrace land and good in the valley/lowland area. Biophysical characteristics directly influenced the crop yield. It is thus called for ecosystem-specific and regionally differentiated research strategy for rice (Barah and Pandey, 2005 and Bhowmick *et al.*, 2005).

Yield Gap Analysis

Since the yield of the hill region varies between lowland and upland and between *jhum* land and valley land, proper weightage has been given during the process of reporting yield on the basis of percentage of area (under different system). Diagnostic surveys have been conducted and some constraints have been identified in few pockets but those studies failed to quantify the magnitude of losses arising from this set of constraints. Information is inadequate which could form the base to prioritise the production constraints in northeastern hill region. The total gap in yield is decomposed into two parts, *viz.*, yield gap I and yield gap II. The yield gap I is the difference between an experiment station's average yield and an on-farm experiment's average yield. It arises from differences in environment that cannot be managed in the farmer's field. Yield gap II, which is the primary concern of the present study. It is the difference between actual farm yield and the yield attained in on-farm experiments. This gap reflects various biotic, abiotic stresses and socioeconomic constraints.

As expected, yield gap in percentage terms in Table 4.1, is smaller in Manipur (85%) where one expects to find modern varieties, high inputs and assured irrigation to a large extent. The highest yield gap in percentage terms occurs in Nagaland (144%), which is grown mainly under stress situation, in marginal lands and is susceptible to a wide range of pests and diseases. The average yields obtained by the farmers under most situations are less than half of the yields at experimental station. Like yield gap I, the yield gap II is also very large. This supports the view that though the HYVs were started with a premise that they would perform superior even under stress condition but since they demand more nutrients, their performance in low fertility condition is inadequate. In-depth socioeconomic survey was carried out in east and west Khasi Hills covering both the mid and high altitude areas as well as valley area of Meghalaya during the year 2004-05 in order to get an idea of yield gap at farm level. Micro-level farm level data (Table 4.2) indicates that average productivity of rice in the mid altitude areas of east and west Khasi Hills of Meghalaya ranged between 500 to 2,500 kg/ha during 2004-05. This wide range of variability may partly be explained by the extreme diversity in rice ecosystem, even within a small patchy pocket of rice growing areas, that possess a major constraint and

Table 4.1

Estimated Yields and Yield Gaps in Rice in NER of India

(kg/ha)

Particulars of Yields	*Manipur*	*Nagaland*	*Meghalaya*	*Mizoram*	*Sikkim*	*Tripura*
1. Experiment station yield*	6775	4578	5343	4933	3745	4555
2. On-farm experiment yield*	4524	3551	3505	2475	2310	2587
3. Actual farm yield **	2442	1456	1875	1527	1435	1329
4. Yield gaps***						
a. Yield gap I	2251	1027	1838	2458	1435	1968
b. Yield gap II	2082	2095	1630	948	1243	1258
c. Yield gap (I+II)	4333	3122	3468	3406	2310	3226
d. Yield gap total (per cent)	177.44	214.42	184.96	223.05	216.49	242.74
e. Yield gap II (per cent)	85.26	143.89	86.93	62.08	82.00	94.66

Source: *Various experimental results from ICAR Research Complex for northeast region—Manipur, Nagaland, Meghalaya, Mizoram, Sikkim and Tripura centre.
**Generated by the study; ***Self-calculated following standardised methodology.

Table 4.2

Exploration of the Potential Productivity (Kg/Ha) Gap of Rice in Mid-Altitude Areas of East and West Khasi Hills of Meghalaya

Ranges of Productivity	*Average Yield (in Kg/ha)*	*Potential Gap*	*Percentage of Farm Households*
500-699	642 (13.3)	1904	11
700-999	876 (13.3)	1670	7
1000-1199	1156 (7.66)	1390	13
1200-1700	1377 (18.24)	1169	24
1701-2000	1733 (20.02)	813	22
2001-2200	2167 (14.26)	379	7
2201-2500	2383 (13.75)	163	7
2501-3000	2546 (5.65)	-	9

Note: Figures in parentheses show coefficient of variation (in per cent).

Source: Primary Survey (2005-06). Umiam, Meghalaya: Division of Agricultural Economics & Statistics, ICAR Research Complex for NEH Region.

therefore demand for seeds are also very much location specific. So, to grow the quality seeds are a big and complicated issue and hence a great challenge for technological intervention. Our analysis shows that in spite of prevailing increasing return to scale (+0.6035) indicating that productivity of rice can be increased if input intensive programme is implemented, but the main difficulty of implementation is the complex and diverse socio-economic background of the farming community in the region. Field information clearly indicates that the average level of chemical fertiliser application in rice field is too low (maximum 24 kg/ha) and in some cases it is insignificant. This may be the main reason to explain the potential yield gap among and across the rice-growing farmers in mid altitude area of Meghalaya and the region as a whole. Table 4.2 clearly indicates that the highest level of achievable (9%) yield is explained mainly by management factors and partly by the biotic and abiotic constraints, calls for the target oriented holistic approach both at scientific level and policy level. At the strategic level, it is possible to bring at least another 14 per cent farmers among the highest yield group.

Table 4.3

Variations of Productivity and Fertiliser Application of the Local Rice Varieties in High Altitude of East Khasi Hills Area of Meghalaya

Varieties	*Number of Observation*	*Average Yield (in kg/ha)*	*Variation of Productivity**	*Average Fertiliser Application(in kg/ha)*
Thaka	5	1380	13.94	24 (40.07)
Thangsaw	19	880	18.67	17 (16.11)
Nonglowai	27	920	9.09	18 (15.21)
Ethati	6	1320	46.22	21 (42.59)
Leit kynsow	9	940	22.06	14 (29.09)
Megha I&II	7	1522	27.08	22 (47.13)

Note: * and figures in parentheses show coefficient of variation in per cent.

Source: Primary Survey (2005-06). Umiam, Meghalaya: Division of Agricultural Economics & Statistics, ICAR Research Complex for NEH Region.

In high altitude area farmers preferred to grow local landraces than HYVs. Among the local varieties, *Nonglowai* is mostly preferred (Table 4.3) even though its productivity (920 kg/ha) is less as compared to other local landraces. The highly differentiated yield rate fluctuation along and across the local landraces is directly reflected by the estimated value of the coefficient of variation. The magnitude of coefficient of yield variation ranging between 46 to 9 per cent indicate the instability and high risk. Farmers are not willing to accept the varieties, as these varieties need some modification in the form of incorporation of two or three desirable traits for the easy adoption of the same.

The actual yield gap between locally used varieties and *Megha* I & II variety is around 563 kg/ha whereas at potential level it is around 321 kg/ha (Table 4.4). But the coefficient of variation is relatively high in case of Megha I & II variety (27.08) than the mostly used local variety (9.09) i.e., *Nonglowai* (See Table 4.3). This variation may be explained due to adaptability and stability of the local variety. In case of high yielding rice variety like Megha I & II, farmers' unwillingness to accept may be interpreted that they are not willing to accept the variety(s) as these varieties are not fulfilling their test. Hence the varietal selection programme should take into account the local needs. This synergistic pattern is a proven path for production stability, which should not be neglected. Moreover, the farming community reports that these varieties

(HYV) are not very long living as they tend to loose their viability and productivity, become increasingly susceptible to pest/disease attack, within few years. With prolific adoption level, the marginal gain in yield is a paradox, which further raises a pertinent question that why the yield of rainfed hilly ecosystem are so low, even under accelerated adoption of modern technology? Thus it requires more insight and serious soul-searching on-farm investigation and to determine the cause and effect relationship (Barah and Pandey, 2005). Farmers' perception should be integrated in the process of technology improvisation programmes.

Table 4.4

Exploration of the Range of Potential Productivity (Kg/Ha) Gap of Rice in High Altitude Area of East Khasi Hills of Meghalaya during 2004-05

Locally used Variety		*Megha I & II*		*Gap*	
Range of Productivity	*Average (kg/ha)*	*Range of Productivity*	*Average (kg/ha)*	*Actual (kg/ha)*	*Potential (kg/ha)*
200-400	325	600-1100	783	458	692
401-600	573	1100-1400	1164	591	311
601-800	750	1401-1500	1388	638	87
801-900	900	1501-1700	1475	575	0
Average	591		1154	563	321
CV	34.52		25.65		

Source: Primary Survey (2005-06). Umiam, Meghalaya: Division of Agricultural Economics & Statistics, ICAR Research Complex for NEH Region.

Regarding the perception of farmers for improved varieties do not appear to have any significant advantages over traditional varieties. For example, even after 12 years of releasing the *Megha* I & II variety, the rate of adoption is only 27 per cent (Datta *et al.*, 2005). Even the process of adoption being very slow calls for participatory mode of breeding programme. Improvement in yield was observed to be most powerful and positive factors influencing the adoption of improved rice variety seeds. This improvement in rice yield was coupled with the requirement of higher amount of inorganic inputs. However, as per the farmers' perception it is explicit from the results that the improved rice variety seeds are more susceptible to pest and diseases as compared to the traditional variety seeds. It is quite notable that better market price was not a significant factor contributing to the adoption of improved rice variety seeds. Rice production in these hilly states is mostly for home consumption and traded through small-scale informal channels with low volume, therefore, the rice market is not at all commercialised, perhaps the reason that improved rice fail to fetch higher prices. Stickiness of cooked rice is another factor that the local people preferred most and, is a common characteristic of traditional rice varieties, however, it was not observed to be a significant factor for non-adoption of improved variety seeds.

Farmers rely mainly on traditional varieties, which are low yielding but seem to be tolerant to some of the stresses that characterise rainfed ecosystems. Farmers are risk averter and prefer to be safe than sorry; they tend to prefer an inferior outcome that is relatively certain to the prospect of a higher average return with a greater degree of risk attached. Further, in terms of characteristics such as cooking quality, aroma, taste, keeping quality of steamed rice and suitability of use as food other than steamed rice, farmers indicate a clean preference for traditional varieties. It needs some modification in the form of incorporation of two or three desirable traits for the easy adoption of the same. Field investigating clearly indicates that most of the local varieties have medicinal properties. For example, the local genotype rice variety *Chakhau paireition* is given to women at advanced stages of pregnancy; *Chahan amubi* is used in mouth ulcers. The impact of low yield of the locally used varieties confirms that shortfall in production has been more acute at the household level than at aggregate level. Therefore, the problem of household food security looms large.

In other words, in the high altitude area, socioeconomic reasons are more dominating for low yield and hindering the productivity growth than in mid altitude and valley area where biotic factors plays a major role for the same. From the above facts it is clear that in this region, the farmers preferred and are continuing to grow a large number of landraces, which are highly eco-friendly where local crop varieties are most suitable to grow because these are more resistant to biotic and abiotic stress in comparison to the modern varieties. Secondly, are we interested to lose those landraces at the cost of enhancing productivity through HYV? The real threat of HYV rice varieties in the region is quite high in terms of its conservation, but we cannot afford to lose any potentially viable landraces, because once a landrace is lost it is lost forever. As the value of one potential landrace is exceedingly high (not possible to measure exactly as it will depend on its future uses), we must conserve it for future utilisation. It is known that the adoption of high yielding, uniform cultivars have led to considerable reduction in the number of genetically viable species used in agriculture. Finally, the non-adoption or slow adoption rate of modern agriculture has helped the farmers of these backward states, to maintain and sustainably manage a huge agro-biodiversity, which has enormous public value for the country. Therefore the opportunity of *in situ* on farm conservation is quite high and might be feasible with least opportunity cost and may be needed to give top priority than for adoption of HYV varieties.

The yield loss indicated by the farmers due to the constraints is occurring beyond the economic threshold level. The constraints, which are causing major production losses, are discussed below. Table 4.5 reveals the fact that the nature and magnitude of losses due to various production constraints varies across states and within the states due to fragility and sloppy land within the states.

Average yield loss in paddy due to all the constraints is presented in Table 4.5. i.e., Nagaland (2,095 kg/ha) followed by Manipur (2,082 kg/ha), Meghalaya (1,630 kg/ha), Tripura (1,258 kg/ha), Mizoram (948 kg/ha) and Sikkim (875 kg/ha). In percentage terms it is about 144, 85, 87 ,95, 62 and 82 per cent respectively of the average farm level yield (Table 4.1). Therefore, there exists a potential range between 144 to 62 per cent to increase the productivity of those selected states in NER through elimination of the losses.

The loss due to technical constraints was ranging between 74 to 65 per cent of the estimated yield gap (Table 4.5). Insects and pests are at the top causing a maximum yield loss in case of Meghalaya (458 kg/ha), Manipur and Nagaland (485 kg/ha) each whereas in case of Mizoram it is (260 kg/ha), Tripura (214 kg/ha) and Sikkim (202 kg/ha) accounting for as high as 33 per cent to 16 per cent of farm level yield. In Mizoram and Sikkim, insects-pest and diseases are the major factors among all the biotic constraints. Constraints within pests & insects, the losses caused by stem borer alone accounts for 155 kg/ha for Manipur followed by 114 kg/ha for Nagaland, for Tripura it is about 58 kg/ha while for Meghalaya it is about 22 kg/ha. It was reported that there was a large-scale paddy-hispa invasion in Tripura during 2001 and the yield loss was around 15 to 20 per cent. Other major pests & insects, in order, are gundhi bug, gall midge, leaf folder, case worm, armyworm and rats. Diseases rank 2nd in the list among all constraints causing a maximum yield loss about 495 kg/ha to 189 kg/ha of those states (Table 4.4). Among diseases, neck blast and leaf blast are the major diseases causing yield losses. In Manipur, the yield loss due to neck blast is about 259 kg/ha whereas in Nagaland it is estimated to be 71 kg/ha and it is as low as 20 kg/ha in case of Meghalaya and Tripura. The yield loss due to leaf blast is ranging between 181 kg/ha to 43 kg/ha among those states. Sheath blight, brown spot and bacterial leaf blight are important diseases of low and mid altitude. In high altitude areas, sheath blight is increasing. Among bacterial diseases, bacterial leaf streak is the most serious problem contributing nearly 58 kg/ha to 24 kg/ha of yield loss. Severe infestation of weeds, mainly aquatic and non-aquatic, during *kharif* season, is an important constraint which together contribute at the tune of 142 kg/ha to 34 kg/ha to the yield losses. The crop also suffers from poor soil health, as a result of nutrient deficiency, low organic matter content, various forms of soil erosion, and zinc deficiency. Together they caused 299 kg/ha to 34 kg/ha yield loss of paddy. The yield losses across states have been summed up in paddy production system in NER region in Table 4.5.

Table 4.5

Estimated per Hectare Yield (in kg) and Income (in Rs.) Loss due to Biotic and Abiotic Constraints of Paddy Production in Northeast Region

Constraints	Manipur	Nagaland	Meghalaya	Mizoram	Sikkim	Tripura
Pest & insects	484.84	484.58	458	260	202	214.13
Diseases	478.89	495.38	207.2	212	189	265.04
Soil & nutrient	298.56	77.96	163.2	34	96	213.85
Weeds	141.83	34.05	62.2	55	52	88.66
Water deficit/excess	39.22	56.17	59.80	20	39	47.38
Cold problem	10.20	183.59	73.81	13	30	14.31
Others	17.58	26.68	35.79	24	42	18.39
A. Loss due to technical constraints (kg/ha)	1470.68 (70.64)	1358.41 (64.84)	1060 (65.03)	618 (65.19)	650 (74.28)	861.76 (68.50)
B. Loss due to socio-economic (kg/ha)	611.32 (29.36)	736.59 (35.16)	570 (34.97)	330 (34.81)	225 (25.72)	396.24 (31.50)
Total yield loss (kg/ha)	2082	2095	1630	948	875	1258
Gross income loss (Rs./ha)	11451	11523	8965	5214	4813	6919
Net income loss (Rs./ha)	13582	13659	10628	6184	5512	8202

Source: Primary Survey (2005-06). Umiam, Meghalaya: Division of Agricultural Economics & Statistics, ICAR Research Complex for NEH Region.

5 Scope of Sustainable Agriculture (Farming System Research) in Northeast Region

The value of agricultural value at the aggregate level may not represent the true picture of the agricultural system in the northeastern states. Therefore before giving any policy implications for the region careful micro-level studies are essential and particularly the long-term viability of various farming systems are to be analysed under different situations. Because many modified/improved farming systems have the desired potentiality to improve the farm income substantially. Keeping in view of farming condition, Farming System Research (FSR) approach, which constituted an important step forward in agricultural research in India particularly in the hilly states of northeast region, was advocated by ICAR Complex for NER region. There is a scientific reason behind it to protect the land and water from degradation caused by adverse effect of *Jhum*/slash-and-burn/shifting cultivation without proper conservation measures in the hilly slopes of the region. Constant effort has been met to develop suitable farming system alternative to shifting cultivation. Some of the programmes being implemented under the auspices, are: soil conservation and land reclamation for permanent agriculture in hills; setting *jhumias* on wet terraced land for growing horticultural crops; engaging shifting cultivators as wage earners in the cash crop plantation and setting them on forest land in small pockets with some provision of basic amenities like schools, sales depot etc. The proposition is that the hill farmers should move to sustainable farming practices which would be a bundle of terraced agriculture and horticulture crops, but so far the research intervention have failed (Chawii, 2001).

To get an outlook of agriculture for NER as such, following questions are to be looked in this framework: first, do we have identified the rationality of *jhum* cultivation and the problems of *jhum* practitioners (*jhumias*) correctly? Second; how far the proposed alternative farming system model is problem solving to the shifting cultivation considering the existing socio-economic condition of the farmers? And, finally; is the present orientation of farming system research appropriate or need based?

To get an in-depth insight of the region, following exercises were carried out which will give the outlook of agriculture for the region as a whole. Database for this section was compiled from various annual reports published in ICAR Research Complex, Barapani, during 1995-2004. Annual reports for the preceding decade were taken into consideration to obtain large number of alternative activities, which have been experimented at the model farming system research. The input use and returns from the livestock unit (here one milch cow) was calculated from the data recorded at farming system research farm. Information on *Jhum* cultivation was used from various published (Singh *et al.*, 2003) and unpublished sources and also from personal communication with the scientists from different disciplines. Expenditure on input and value of output was calculated based on the cost and price prevailing at 2004-05 current prices.

Three different situations were simulated based on the primary objective of farming with various assumptions. The situations are explained below:

Situation I: Availability of fixed and operational capital is abundant, area under agriculture, horticulture and forest block is fixed, area under different crops is interchangeable and growing rice is not an essential activity in this situation. This simulation was assumed to resemble the situation when the primary objective of farming is profitability and possess highest flexibility among the three alternative situations. This scenario has been designated as progressive farming.

Situation II: Availability of capital is not restricted, area under agriculture, horticulture and forest block is fixed, area under different crops is interchangeable and growing rice is an essential activity in this situation. This simulation is a blending of some flexibility (crop-area interchangeable depending on their profitability) and rigidity (area under rice is a must). So, this plan resembles the situation when the objective of farming is profitability as well as to maintain household food security. This scenario has been designated as semi-progressive farming.

Situation III: Availability of capital is not restricted, area under agriculture, horticulture and forest block is fixed, a large number of crop activities are essential and minimum area under these crops are restricted, and some rice area is must in this model. This is the situation mostly prevailing in NER states, where large number of crop activities are taken in a piece of land i.e., mixed cropping to fulfill their households' food requirement. The profitability is not the primary objective of farming and the whole system is operating at subsistence level.

Area allocation of optimum crop plans under alternative situations were estimated along with their corresponding net returns (Table 5.1), results clearly indicated that as the large number of activities are included in the model (Situation III), the profitability declines (Rs. 5,138 per ha).

Table 5.1

Area Allocation and Net Returns of Optimum Crop Plans under Alternative Situations in Agri-Horti-Silvi-Pastoral Farming System Model

Activities	*Area Allocation (in per cent)*			
	Situation I	*Situation IIa*	*Situation IIb*	*Situation III*
Ginger	16.50	12.62	17.48	6.80
Radish	15.53	-	-	-
Assam lemon	32.04	32.04	32.04	10.68
Broom grass+forest	35.93	35.92	35.92	25.24
Rice	-	19.42	14.56	7.77
Maize	-	-	-	11.65
French bean	-	-	-	6.80
Orange	-	-	-	9.71
Pineapple	-	-	-	9.70
Guinea grass	-	-	-	11.65
All	100	100	100	100
Net returns (Rs./ha)	12784	6460	7534	5138

The reason is high diversification which includes some comparatively less profitable activities. Situation I give highest net return (Rs. 12,784 per ha), where the choice of crops purely depends on the magnitude of profitability of individual crops. The crop activities in this model include ginger (17%), radish (16%), Assam lemon (32%) and broom grass along with forest (36%). Broom grass (provides green fodder and inflorescence is used for making broom) is an attractive economic activity prevailing in this area, which does not require any input cost. In horticultural block Assam lemon was found in the optimum plan in lieu of orange, Khasi mandarin, guava and pineapple. The productivity of orange in this region has declined drastically and performance of guava and pineapple crop was observed to be poor, hence Assam lemon emerged as the most profitable activity among horticultural crops. Ginger is the main cash crop and quite profitable followed by radish, which were included in the final solution of the optimum plan.

Situation II (resembling semi-progressive status) was further analysed into two alternative situations (Situation IIa and Situation IIb) to examine the change in profitability due to change in area under rice. The results showed that area under rice has inverse relationship with net return. In Situation IIb area under rice (14.56%) declined by nearly 5 per cent as compared to Situation IIa (19.42%), which improves the net return per hectare by 17 per cent (i.e., from Rs. 6,460 to Rs. 7,534). Rice has not been considered as an important crop activity in agri-horti-silvi-pastoral system because of its low profitability resulted from poor yield performance (1.00 to 1.20 t/ha). The idea was to replace the rice area (upland) with some alternative remunerative crop to ensure higher profit earnings and to drive the farming operation at higher level i.e., to commercial scale. But the ground reality is quite different, because, farmers (*jhumias*) prefer to grow rice to fulfill their household food requirement and also due to strong preference for locally grown glutinous rice and emotional attachment. Mostly the *Jhum* areas that are well connected by road linkages are operating at this semi-progressive stage.

Situation III which was hypothesised to represent typical subsistence farming shows lowest net return (Rs. 5,138 per ha) as compared to all other situations (Table 5.1). Here the only purpose of farming operation is to fulfill their household food demand. The farming includes a large number of diversified agricultural, horticultural and silvi-pastoral activities and profitability is least considered. This situation is prevailing in remote areas where the market-driven forces are almost absent or weak. So, the Situation III represents subsistence farming with large number of crop activities (mostly mixed cropping) with low net returns, but this farming engaged maximum number of labour forces (though not gainfully employed) and satisfies most of their household requirements.

The profitability of existing *Jhum* cultivation system is as comparable to the proposed permanent agriculture system (agri-horti-silvi-pastoral) when operating at semi-progressive state. Further, as per our assumption, semi-progressive farming is fulfilling the dual purpose of farming i.e., household food security as well as some earning. The findings clearly indicate that if the existing farming system is fulfilling this dual purpose, then why the *jhumias* will not intend to adopt the permanent agriculture system? Moreover, the decisions on adoption are influenced not only by the positive attributes of the technology, but by a host of others factors like resource endowments, risk preference, family requirements, resource flows, the price signals, product and factor market conditions, etc. Social scientists argued that improved technologies were often rejected because they failed to mesh with this real life condition under which the farmers operate, which calls attention for a research that is need based and address these constraints (Jha, 2001).

Table 5.2

Improvement in Net Returns from Proposed Agri-Horti-Silvi-Pastoral Farming System Model Over Shifting Cultivation with and without Livestock Component

Situations	*Net Returns from Agri-Horti-Silvi-Pastoral Model (Rs./ha)*		*Changes in Net Return (Rs./ha)**	
	WoLIV	*WLIV*	*WoLIV*	*WLIV*
Situation I	12784	17684	+5547 (76.65)	+10447 (144.00)
Situation II	7534	12434	+297 (4.10)	+5197 (71.81)
Situation III	5138	10038	-2099 (29.00)	+2801 (38.70)

Notes: WLIV and WoLIV represents the livestock and without livestock component, respectively; *Average net return from shifting cultivation was calculated to be Rs. 7,237 per ha. Figures in parentheses show change in percentage.

Interestingly inclusion of dairy based livestock component which is an integral part of farming system, shows that it improved the net return of the agri-horti-silvi-pastoral farming system substantially (Table 5.2) (144%, 72% and 39% under Situation I, II and III, respectively) irrespective of situations under consideration. But in reality, *jhumias* rear pig, which is an integrated part of their farming system rather than dairy based livestock. It clearly indicates that inclusion of piggery along with dairy will further enhance the households' family income apart from fulfilling their own consumption demand.

Table 5.3

Financial Feasibility of Agri-Horti-Silvi-Pastoral Farming System Model vis-à-vis Shifting Cultivation

Situations	*NPV (in Rs.)*		*BCR*		*IRR (in per cent)*	
	WLIV	*WoLIV*	*WLIV*	*WoLIV*	*WLIV*	*WoLIV*
Situation I	53833	37024	1.32	1.20	46	37
Situation II	24092	11270	1.21	1.10	30	22
Situation III	12953	1747	1.16	1.02	24	15
Shifting cultivation	1265	-	1.07	-	16	-

Notes: WLIV and WoLIV represents livestock and without livestock component, respectively.
NPV - net present value; BCR - benefit cost ratio, and IRR - internal rate of return.

Considering the low economic status and poor resource endowment with the *Jhum* cultivators, financial feasibility study reveals that shifting cultivation cannot be termed as an absolutely non-viable option for investment. So, the farming system research should be focussed on improvement of shifting cultivation rather than complete weaning it out. This view can be refreshed by the observation made by Mishra and Gupta (1998), stating that, ICAR Complex has designed and demonstrated several cropping systems for different agroclimatic zones along with input management and ensuring food security as alternatives to *Jhum* but with little success. Perhaps the alternative solutions that were offered to *jhum*, ironically were those formulated in labs and in scientists' pilot plots designed more to test their assumptions than to reflect the ground reality in the *jhumia's* field. It is not mere a perception that the scientific research on shifting cultivation has been utterly inadequate not only in terms of insights into the prolonged so-called 'primitive' farming methods used by many hill communities, but also in finding out alternative land use packages to improve the *jhumias* farming method (Chakraborty, 2005). Experts have now started thinking about modification of shifting agriculture instead of completely replacing them. Financial feasibility study also investigates the change in profitability due to incorporation of livestock component in the farming system model. Result (Table 5.3) depicts that magnitude of all the investment criteria (NPV, BCR and IRR) improves substantially due to livestock component, irrespective of situations under consideration. This again reconfirms the importance of livestock component in the farming system models.

Prioritisation of Income Components under Shifting Cultivation Practices

The importance of livestock component to improve profitability of farming system has already been examined and discussed in earlier section. Present decomposition analysis was carried out to prioritise the different income sources to enhance the returns from shifting cultivation in equitable manner. The result is expected to help in prioritising the incorporation of different income sources to achieve higher social benefit and higher distributive justice. Gini index from total household income was estimated to be 0.204 at Umroi area indicating that income distribution among shifting cultivators are almost even. This implied that, in one hand the shifting cultivators are constrained with low returns, but on the other hand the positive aspect is the equitable distribution of

Table 5.4

Decomposition of Source-wise Farm Income Inequality in Umroi (Meghalaya)

Sources of Income	*Correlation with Rank of Total Income*	*Gini of Sources*	*Income Share*	*Source of Income Inequality*	*Per cent Contribution to Total Inequality*	*Marginal Effect*
Agriculture	0.841	0.375	0.356	0.113	0.554	+0.197
Livestock	0.921	0.183	0.064	0.011	0.054	-0.010
On & off-farm employment	0.712	0.163	0.296	0.034	0.167	-0.013
Business	0.933	0.245	0.167	0.038	0.187	+0.020
Others	0.634	0.102	0.116	0.008	0.039	-0.077

income. Results (Table 5.4) showed that the share of agricultural income was highest (36%), followed by income from on and off-farm employment (30%), business (17%), others (17%) and livestock (6%) (Table 5.4). It is interesting to note that income from on and off-farm employment is quite high (30%) indicating the alternative source of income of the farmers. But the real concern is, even having the alternative income sources, merely 28 per cent of the sample farmers were practising solely shifting cultivation and rest were following shifting (part of their land) as well as permanent agriculture. The possible reason might be firstly, farmers are having special attachments with the traditional practice as it is not only fulfilling their typical food grains requirements but also it is their way of life. Secondly, shifting cultivation does not require any intensive input management practices, which is absolutely matching with their low resource base endowment factor. Third, abundance of land (though land-man ratio is squeezing but still it is not a crisis in hilly terrain as far as shifting cultivators view is concerned) and finally, it is a low-input-low-output but highly labour intensive, suitable for this environmentally fragile region, where agricultural operations are bound to suffer from high weather risk.

Results on percentage contribution to total inequality show, agricultural income alone is contributing nearly half of the total income inequality (55%) and also the marginal effect is positive. The positive marginal effect implied that as the agricultural income increases the income inequality will also increase which is not socially desirable as far as social benefit is concerned. Similar is the case for income from business also. But the marginal effect was observed to be negative for other income sources, namely, income from livestock, on and off-farm employment and others income. The negative marginal effect suggested that additional income from a particular source (say, livestock) will reduce the total income inequality, which is socially desirable. This re-emphasises the importance of livestock component in farming system for its double impact, improvement in profitability (desirable from private point of view) as well as reduction of household income inequality (desirable from social point of view), thus, strongly suggest incorporation of this component will help in the process of development in equitable manner. Along with the livestock component, results also suggested opening up avenues for on and off-farm income sources as well as income from other sources.

6 Risk and Uncertainty

Revitalising the agriculture sector is a priority for overall development and agricultural credit is the most crucial factor to make the agrarian economy vibrant, particularly in the areas where the benefits of green revolution have not yet percolated, like the northeast region of India. Realising the need to enhance credit flow to agriculture, the Government of India, in consultation with the Reserve Bank of India (RBI), National Bank for Agriculture and Rural Development (NABARD) and commercial banks, announced on June 18, 2004, a credit package for agriculture sector, which envisaged doubling of agricultural credit over three years. To step up agricultural credit from all lending institutions, estimated at Rs. 80,000 crore during 2003-04, by 30 per cent to around 1,50,000 crore during 2004-05 with the share of various agencies at Rs. 57,000 crore by commercial banks, Rs. 8,500 crore by regional rural banks (RRBs) and Rs. 39,000 crore by cooperative banks (Government of India, 2005). Despite the continuous efforts, the flow of credit to the agricultural sector failed to exhibit any appreciable improvement mainly due to the fact that commercial banks are not tuned to the needs and requirement of the small and marginal farmers, while cooperative, on the other hand, lack resources to meet the expected demand (Mohan, 2004). In fact, total amount outflow of funds from the Centre to this region is estimated to be about Rs. 12,000 crore (US $2.7 billion), which is higher than the disbursements received by India as a whole from multilateral institutions like the World Bank and Asian Development Bank. It is also higher than the total aid received by Bangladesh. Thus, the flow of funds is not the problem, but they need to be used in such a way that the seeds of sustainable development are sown so that the growth of the region becomes self-propagating (Mohan, 2003). However, the agricultural advances in this region are quite low as compared to other parts of India. For example, the share of agricultural advances by all banks in NER was as low as 11.75 per cent in Assam followed by 14.66 per cent in Nagaland, 12.35 per cent in Manipur, 20 per cent in Arunachal

Pradesh and 26.28 per cent in Tripura (Government of India, 2000). Thus, the NER on one side is receiving substantial amount of financial resources but the agricultural sector, so far has failed to attract adequate amount of investment from institutional sources. The nationalised banks have shown a lukewarm response towards financing crop loans particularly in view of the paucity of updated land records, lack of entrepreneurship development, loan repayment status of land poor, which increase the amount of non-performing assets (NPAs) in agriculture for commercial banks and in turn increase the risk in agricultural financing.

On the other hand the farmer in this region is mostly following the traditional agricultural practices, which is characterised by 'low-input-low-output' situation. The demand for agriculture loan is expected to be induced by the availability and adoptability of suitable modern agro-technologies. But the advancement of modern agrotechnologies in this region is very slow and severely constrained by topographical features (70 per cent is hilly with 42 to 26 per cent forest area of each state). The modern agrotechniques, such as high yielding varities (HYVs) are usually required high input use (fertiliser, pesticide and irrigation etc.), in which lumpy investment is needed, unfavourable to the existing socioeconomic condition of the hilly farmers. Most importantly, the higher yield for improved crop varieties is also associated with high yield variability, under these typically sloppy/hilly areas which is bound to increase the risk and uncertainties of farm income, is the major impediment for agricultural development in NER.

The analysis of credit availability indicates that direct short-term credit per ha in the northeastern India is less than Rs. 100 against a national average of Rs. 422 per ha. This is a meagre amount in the context of modern agricultural technology, which is basically a capital-intensive activity. While credit availability per hectare in Kerala is more than Rs. 5,000, in Bihar it is only Rs. 50 and in the northeastern states as a whole about Rs. 25. Statewise credit-deposit ratio (Table 6.1) clearly indicates that in case of Arunachal Pradesh (AP), till 1998 only 11-13 per cent of the deposited amount was invested in the states whereas 1999 onwards it is moderately on a higher (32.61%) side as compared to earlier period. In case of Assam, it is in reverse direction as it declined in 1999 to the extent of 19.25 per cent whereas in the earlier period it was in the range of 33-39 per cent. In case of Manipur, the trend of credit deposit ratio is slightly declining (47.60%) in 1999 as compared to earlier periods where it was ranged between 54-59 per cent during 1995-1998. At least compared to other NER states, the condition of credit deposit ratio (C-D) in Manipur is favourable and reasonably comparable with all-India figures, which was ranging between 55-62 per cent during the same period. The worst state within NER in terms of C-D ratio is Meghalaya where it ranges between 14-16 per cent. Overall in the NER, the trends of C-D ratio are declining from 35.6 per cent to 30.29 per cent during 1995 to 1999.

Table 6.1

Credit–Deposit Ratio of Scheduled Commercial Bank in Northeast Region

State	*1995*	*1996*	*1997*	*1998*	*1999*
Arunachal Pradesh	12.4	10.4	10.7	13.1	32.61
Assam	38.7	39.3	35.2	32.9	19.25
Manipur	58.2	53.7	57.5	58.8	47.60
Meghalaya	15.8	14.4	14.3	15.2	16.68
Mizoram	16.5	16.2	14.0	23.2	23.40
Nagaland	37.8	27.4	24.0	18.3	21.25
Tripura	47.5	42.1	37.7	34.0	35.00
Northeast region	35.6	34.5	31.2	29.9	30.29
All India	55.6	61.9	57.3	55.5	

Source: GoI (2000).

A Case Study on Risk and Uncertainties

The rice is cultivated in three distinctly different situation in the hilly terrain of Meghalaya, which are: (i) rice in *jhum* area (i.e., shifting or slash and burn method), (ii) rice in upland area (in sloppy or undulated hillside, terrace area, not primitive as like in *jhum* cultivation), and (iii) rice in valley area (between hills and most fertile land). The magnitude of coefficient of variation (CV) in total return, net return and rate earned on equity capital showed that the uncertainties of these variables are quite high under *jhum* as well as in upland condition and the uncertainties have direct relationship with improved method of cultivation (Table 6.2 a and b). The CV was estimated to be very high for various parameters under rice cultivation in valley area. It is evident from the table that the profitability in terms of rate earned on equity capital is realised quite high under

Table 6.2(a)

Risk and Uncertainties in Profitability of Rice Cultivation in Meghalaya under Different Situations

Particulars	*Rice Cultivation in*		
	Jhum Area (n = 55)	*Upland Area (n = 20)*	*Valley Area (n = 25)*
Total cost (Rs./ha)	15181 (37.85)	16870 (45.05)	18338 (47.23)
Total return (Rs./ha)	21883 (45.98)	25513 (50.41)	39554 (52.43)
Net return (Rs./ha)	6702 (38.20)	8643 (53.22)	21216 (49.05)
Yield (Kg/ha)	1614 (48.48)	2500 (45.63)	3057 (46.29)
Rate earned on equity capital (per cent)	44.15 (62.88)	51.23 (44.20)	116 (53.62)
Unit cost of production (Rs./kg)	9.40	6.75	5.99
Minimum support price (Rs./kg)[$]			
2002-03	5.50	5.50	5.50
2003-04	5.80	5.80	5.80

Note: Figures in parentheses shows coefficient of variation (CV) in per cent.
[$] MSP for common rice.

Source: Estimated based on Primary Survey Data (2005-06). Umiam, Meghalaya: Division of Agricultural Economics & Statistics, ICAR Research Complex for NEH Region.

modern method of cultivation but at the cost of higher uncertainties. This can be explained as: first, the *jhum* cultivation is characterised by low-input-low-output agricultural practice and quite suitable for this hilly terrain, which has been developed from the natural wisdom and from the experience of thousands of years. Second, *jhum* cultivation practice is heavily dependent on the single input, labour (accounts for 80-85 per cent of total cost) which is locally available, but in contrast, improved method of cultivation requires modern varieties of seeds, fertiliser, pesticide and irrigation, and the access of these inputs are severely constrained by availability (like suitable HYV seeds) as well as improper delivery system. Third, like the input delivery system also the lack of output delivery system, i.e., marketing of the surplus produce is a major impediment contributing to the uncertainties of hill farming. Additionally, the minimum support price (MSP) scheme is totally ineffective in this region in view of the higher unit cost of production as compared to the announced

MSP (Table 6.2a). The agricultural farming in hilly region is completely different from the plains as the hill farmers cannot enjoy scale economy due to limited scope of substitutability among the factors of production and also the opportunity of mechanisation is limited, thereby scope for reduction in unit cost is also restricted. To reduce the risk in hill farming and to promote the modern agrotechnologies, government should streamline the appropriate input as well as output delivery system through suitable public support system. Being the MSP declared by the Government of India (GoI) ineffective in this region, a separate price commission can be thought of for this region to encourage agricultural production.

Ginger is the main cash crop for Meghalaya and highly suitable for this region which has good demand for domestic as well as international market. It is cultivated in sloppy track through indigeneous method of practices (making bund, a raised bed along the slope) and subject to very high uncertainties (CV) in total return (71%), net return (68%), yield (72%) and rate earned on equity capital (75%) (Table 6.2b). In spite of high uncertainties ginger is continuing to be a highly preferred cash crop for this region, because: (i) it helps the farmers through supplementing the day-to-day cash requirement throughout the year, (ii) it is easy to store underground, and (iii) it is not perishable in nature as like other vegetable crops, such as tomato, brinjal, capsicum etc. The unit cost of production is Rs. 5.44 per kg only as compared to the retail price, that ranges from Rs. 40-60 per kg at domestic market. Clearly, ginger has the comparative

Table 6.2(b)

Risk and Uncertainties of Ginger Cultivation in Ri-Bhoi District of Meghalaya during 2002-03

Sources of Risk	*Average*	*Coefficient of Variation (in per cent)*
Total cost (Rs./ha)	27246	76.25
Total returns (Rs./ha)	45072	71.46
Net returns (Rs./ha)	17826	68.64
Yield (kg/ha)	5008	72.77
Rate earned on equity capital (per cent)	65	75.75
Unit cost of production (Rs./kg)	5.44	-

Note: No minimum support price (MSP) is announced for ginger.

Source: Estimated based on Primary Survey Data (2005-06). Umiam, Meghalaya: Division of Agricultural Economics & Statistics, ICAR Research Complex for NEH Region.

advantage to be grown in this region. So, in spite of high instability in production and profitability, it has continued to be the most preferred cash crop. Several steps are necessary to strengthen backward linkages and standards as the global market demands. First as the bulk of our production is consumed domestically, there is urgent need to improve domestic standards. Second, although the Spices Board is encouraging use of logo and brands through certain schemes, there is need for more innovative schemes to speed up this process. Third, as the standards differ significantly between professional corporate bodies and proprietary concerns, government policy must encourage the former. Finally, the corporate sector has to seriously think in terms of paying farmers partially in terms of bonus share (Datta *et al.*, 2002).

Relationship of Profitability and Farm Size

Quadratic form of production function was employed to explain the relationship between profitability (rate earned on equity capital) and farm size (cropped area under ginger or rice) (Table 6.3). The specified function was found to be fit only for ginger and overall 29 per cent of the total variation in profitability was explained by the variation in cropped area. The estimated parameters indicated that as the cropped area under ginger cultivation increases, the rate of return earned on capital also increases, but subsequently starts declining after some critical cropped area. This can be

Table 6.3

Quadratic Production Function of Responsiveness of Profitability to Cropped Area for Ginger and Rice Cultivation in Meghalaya

Variables	*Rate Earned on Equity Capital for*	
	Ginger	*Jhum Rice*
Intercept	1.3584***	0.6963
	(0.4131)	(0.4137)
X (cropped area)	0.6839*	- 0.1488
	(0.4301)	(0.9806)
X^2 (cropped area)	- 0.2680**	- 0.0386
	(0.1368)	(0.4765)
R^2	28.74	19.24
No. of observations (n)	55	20

Note: ***, **, * significant at 1 per cent, 5 per cent and 10 per cent, respectively.

Source: Estimated based on Primary Survey Data (2005-06). Umiam, Meghalaya: Division of Agricultural Economics & Statistics, ICAR Research Complex for NEH Region.

explained as follows: ginger is mainly cultivated through traditional practices and subject to wide yield uncertainty and in turn profitability. As the cropped area increases the yield uncertainties also increases because of different fertility status at varying slopes in hilly track. Additionally, the crop management, particularly intercultural operation becomes difficult. The human labour is the most important factor of production accounting for more than 60-65 per cent of total cost of cultivation. The size of available family labour determines the area under ginger to be cultivated. The yield uncertainty leads to higher risk in terms of lowering the rate earned on capital investment for ginger cultivation. Moreover, the price elasticity of demand for ginger is expected to be very low because the share of consumers' expenditure for ginger in total consumer basket is negligible. The producer's share in consumer rupee has been low meaning thereby that most value addition takes place beyond the farm gate. All these weaknesses need to be removed for making the ginger economy viable and buoyant (Datta *et al.*, 2002).

Under high uncertainty situation, the risk of investing additional amount of non-equity capital also increases considerably and expansion of area under particular crop becomes difficult. The tendency for risk to become greater at an increasing rate as the relative amount of non-equity capital used in a business expands is referred by Kalecki as the 'principle of increasing risk' (Pandey, 1990). This principle can be understood from the results reported in Table 6.4, where the rate earned on equity capital has been calculated under different leverages. The average cost of cultivation of ginger on the sample farmers has been calculated to be Rs. 27,246 per ha. As per the banking guideline the scale of finance for ginger cultivation is Rs. 34,185 out of which Rs. 25,789 is disbursed as kind and Rs. 8,400 as cash component. Since the sample farmers were following the traditional method of ginger cultivation, they were entitled for the cash component (Rs. 8,400) only. Let us assume that farmers have obtained loan of Rs. 5,000 or Rs. 8,000 for ginger and this additional capital amount (non-equity) is being invested for its cultivation along with his equity capital (Rs. 27,246). With full equity in the farm business, both rates of return and rate of loss (may arise due to heavy rainfall, heavy fungal disease infestation or hailstorm) from the investment are 25 per cent because the cost of non-equity capital is zero. However, the situation changes when non-equity

Table 6.4

Estimation of Risk for Capital Investment in Ginger Cultivation in Meghalaya under Different Leverage Ratios

Particulars	Leverages (Non-Equity/Equity Capital)		
	0.00	0.19	0.30
Amount of equity capital used (Rs.)	27246	27246	27246
Amount of non-equity capital used (Rs.)	0	5000	8000
Total capital used (Rs.)	27246	32246	35246
Profitability when rate of return on investment is + 25%			
Gross return to total capital (Rs.)	6812	8062	8812
Cost on non-equity capital @ 14% (Rs.)	0	700	1120
Net return on capital used (Rs.)	6812	7362	7692
Rate earned on equity capital (%)	25.00	27.02	28.23
Profitability when rate of return on investment is - 25%			
Gross loss to total capital (Rs.)	-6812	-8062	-8812
Cost on non-equity capital @ 14% (Rs.)	0	-700	-1120
Total loss on capital used (Rs.)	-6812	-8762	-9932
Rate of loss on equity capital (%)	-25.00	-32.16	-36.45

Source: Estimated based on Primary Survey Data (2005-06). Umiam, Meghalaya: Division of Agricultural Economics & Statistics, ICAR Research Complex for NEH Region.

capital is also used in the farm business. With Rs. 5,000 as non-equity capital, the rate earned on equity capital is only 27 per cent while the loss is 32 per cent. Thus, with the use of non-equity capital along with equity capital the return to equity capital increases by only two per cent (i.e., from 25 per cent to 27 per cent), but under adverse condition it has seven per cent loss. As the leverage ratio increases to 0.30, the return to equity capital under favourable condition is 28 per cent while the loss under adverse condition is 36 per cent. So, the financial risk for equity capital increases along with the leverage ratio under adverse or unfavourable condition. Thus, (a) as leverage increases, there is a tendency for the spread between possible gain and losses to increase; (b) with an equal percentage gain or loss on total capital used in the business the magnitude of loss is greater than the magnitude of gain on owner's equity. Therefore, to increase the use of non-equity capital in farming business, agrotechnologies like HYVs suitable for local condition, needs to be developed. Besides, appropriate and functioning crop insurance scheme should be available to mitigate the risk and uncertainties.

Uncertainties due to Market Imperfection

In earlier sections the analysis of risk in financing hill agriculture (equity as well as non-equity) was concerned with factor of production only. The uncertainty at product market is also equally important which arises due to market imperfection in this hilly region. Highly undulated topography and lack of easy transportation linkages are the major impediments for free mobility of marketed surplus. Several factors can be attributed for this imperfect marketing situation for this region, such as, first, mostly the farmers produces at subsistence level where marketable surplus for the individual farm is very meagre. Second, the opportunity of developing entrepreneurship in market services is very limited due to this low marketable surplus. Third, the transportation cost is too high and the product from this region hardly fetch the competitive price in mainland market. But the region has enormous potential for the products like, off-season vegetables, ginger, pineapple, citrus etc., which urgently demands more focus. Value addition certainly will fetch good return from these crops. The region needs proper public support mechanism to procure and marketing of these products.

Development of market or creation of demand for certain product is not an easy task because the size of population is the most important factor to determine the market demand. Moreover markets are generally embedded in human interaction. The humans that are involved in markets are boundedly rational, they are complex, and culture, history and social forces shape them. An emerging body of work is showing that social capital and what we call 'trust' matters for economic outcomes. Second, the costliness of market interaction and the cost of coordination is a limiting constraint on market performance. Third, market interaction requires enforcement. Not only are human boundedly rational, but they are also opportunistic. Finally, markets require human capital and infrastructure, so capacity building is essential. The northeast region is characterised by very low population density, ranging from highest in Assam (340 per sq km) and lowest in Arunachal Pradesh (13 per sq km). The other states also have very low population density, such as, Manipur (107 per sq km), Meghalaya (103 per sq km), Mizoram (42 per sq km), Nagaland (120 per sq km) and Sikkim (76 per sq km). Under liberalised economy our expectation is that the market will lead to agricultural growth. The fundamental question is whether and how we can create market that will, in fact, deliver what we expect from them. In a market economy we are starting with markets rather than ending with markets, and that is the paradigm shift that we are embarked upon. More is expected of markets than ever before for food security and for agricultural growth. Where growth must be demand-driven in the changing global agro-food system, markets must enable producers to respond to these growth opportunities. We are able to disseminate technologies that are very market dependent and enable us to achieve highly productive agricultural transformation. If we do not have input markets that work, let us move into low-external-input agriculture; let us move into conservation farming, where market is less effective and responses are different. Of course, the other response occurs when output side is functioning, will get price collapses and regress into low-productivity agriculture like NER agriculture.

7 Scope and Opportunity for New Investment

Several opportunities are lying and new approaches are needed to promote the region to take advantage of its own strengths and to turn some perceived weaknesses into assets. In fact the region can be termed as the 'sleeping giant' for agricultural development. Some of the opportunities that can be pointed out are: (i) developing the region as the producer of organic food products which have high demand in international market, (ii) establishing light industries such as, food processing, floriculture, sericulture etc., (iii) encouraging allied agricultural activities, like silk fabrics (silkworm, mulberry, *tussar*, *muga*, *eri*) fishery, particularly ornamental fish rearing, (iii) traditional handloom weaving, (iv) forest based industries, like, collection or growing of medicinal and aromatic plants (at least 120 known medicinal plants are known to grow wild in this region), and (v) there are prospects for essentially oil based industries, such as *agarwood* oil, lemon grass oil, cinnamon leaf oil, citronella oil and like (Mohan, 2003). The financial viability of ornamental fish rearing and opportunity of organic food products are reported in the subsequent sections.

Investment Opportunity: A Case of Ornamental Fish Rearing

The NER produces the bulk of India's ornamental fish exports but remains unattended for developing market opportunities. Several agribusiness opportunities can be explored through systematic collection and selling of native ornamental fish (wild catch) as well as rearing of exotic ornamental fish species (captive breeding) suitable in the region. Business opportunities can be realised at every stage, namely, production, marketing and conservation of ornamental fishes. The financial viability of ornamental fish rearing, gold fish (*Carassius auratus auratus*) is studied in detail and the results are reported in Table 7.1 The financial viability was estimated under various risk and uncertainty situations to examine the sensitivity of the financial viability of the investment. The total number of fingerlings production from the unit has been calculated to be 36,000 per annum and the average price received was observed to be Rs. 6 per piece. The analysis was carried out under four different risk/uncertainty situation, namely: (i) without any risk or uncertainty, (ii) with 10 per cent disease mortality risk of fingerlings, (iii) with uncertainty of 50 per cent fall in market price (from Rs. 6 to Rs. 3 per piece) of fingerlings, and (iv) with combined risk and uncertainty (10 per cent disease mortality and 50 per cent fall in market price). The results showed that the net present value (NPV) was ranging from Rs. 6,37,082 when there is no risk (Situation I) and Rs. 1,21,618 when there is existing risk as well as uncertainty in the business (Situation IV). Similarly, internal rate of return (IRR) was estimated to be 111 per cent and 35 per cent, respectively, under Situation I and Situation IV. The financial analysis clearly indicated that the project is not only viable but also an attractive investment proposal. Despite high risk and uncertainty situation the project will still remain viable.

Table 7.1

*Financial Viability of Gold Fish (*Carassius *Auratus Auratus) Rearing under Different Risk and Uncertainty Situations in Meghalaya (At 2004-05 Constant Prices)*

Investment Criteria	*Without Any Risk or Uncertainty (Situation I)*	*With 10 % Disease Mortality Risk of Fingerlings (Situation II)*	*With Uncertainty of 50% Fall in Market Price of Fingerlings (Situation III)*	*With Combined Risk and Uncertainty (Situation IV)*
Net present value (NPV in Rs.)	637082	543361	168479	121618
Internal rate of return (IRR in %)	111	98	43	35
Benefit cost ratio	3.11	2.79	1.57	1.40

Source: Survey Data (2005-06). Umiam, Meghalaya: Division of Agricultural Economics & Statistics, ICAR Research Complex for NEH Region.

Despite having tremendous growth potential, the ornamental fish sector in the northeastern states remains untapped mainly due to lack of systematic marketing, poor infrastructure (essential for transportation of vulnerable live fishes) and lack of access to market information. The marketing system was highly unorganised and no direct export was being done. Only a few traders collect the native ornamental fishes through local collectors and supply them to different exporters based in Kolkata, Howrah, Mumbai, Chennai, Trivandrum and Cochin. The most prevailing marketing channel of ornamental fish marketing was: collectors–unregistered small traders–wholesalers–exporters at various ports. Estimation showed that exporter enjoyed the lion's share of profit (45%), followed by wholesaler (30%), unorganised trader (20%), and a meagre share of 5 per cent was realised by the collector (Mahapatra *et al.*, 2006). The agribusiness opportunities of these live fishes have been recognised by the producers, collectors and traders. Since, almost the entire volume of ornamental fishes is collected from the natural aquatic resources of the northeastern region, there is a big sustainability threat to these resources. This calls for an urgent attention on formulating sound ecological and economic strategies. It has been stressed in this article that agribusiness opportunities in ornamental fish farming can be realised at every stage, *viz.*, production, marketing and conservation of ornamental fishes. The ornamental fish production in the northeastern region has been observed to be financially as well as economically viable and investment-friendly. With some initiatives by the government like providing incentive to establish ornamental fish production unit, considerable private investment can be attracted to this industry, which would generate additional employment opportunities. With the concerted efforts of all stakeholders, the ornamental fisheries can be developed substantially in the region, which in turn, will gain a larger share in the world market. Public-private partnership (PPP) can be encouraged through establishment of ornamental fish production units in different parts of the region to make this sector more vibrant and remunerative.

Organic Farming: New Opportunity for Investment

The northeastern region is probably the most suitable area of India for organic agricultural farming because of its number of advantages. This region is said to be, by and large, organic by default or the farmers of this region are *de facto* organic producers. The farmers of this region are using very insignificant amount of inorganic inputs, fertiliser and pesticides. Almost in all states of this region, farmers are practising traditional agriculture since thousands of years. Recently, in few pockets of the region, the use of inorganic inputs has been started but the level of use is almost negligible (except Manipur). This low input use status (inorganic input) is the biggest strength for the possibility of organic agriculture in this hilly region. The conversion period of organic from inorganic agriculture is to be much lower and easier in this region as compared to other parts of the country. Moreover, the conversion process will be highly supported by the readily available huge amount of forest biomass and the residues of diverse livestock population. Experiments show that during the transition period, the farmers face the problem of yield losses which will also be minimum in this region as the farmers are using very low level of inorganic inputs. Usually the organic products fetch higher prices, nearly 20-25 per cent premium as compared to the conventional products, which, may compensate the problem of yield losses during early period of conversion and expected to increase the level of income. The NER is having a typical hilly geographical terrain, and the shifting of the produce takes long time to reach even in the domestic markets. As the international market (particularly USA and European Union) is a major buyer of the organic produce, the supply of perishable product from this region will be one of the difficult tasks. So, the government should give much focus on the domestic market to create the demand for organic produce to facilitate and secure the selling of the surplus production from this region. Behavioural study of consumer awareness and preference towards the organic product is very much essential.

Considering these advantages recently the region has been identified as the most potential area for organic agriculture in the country and accordingly, policy thrust has been laid for necessary internalisation of the region. Being the 'hot spot' (one of the 18th mega biodiversity centres of the world) in terms of biodiversity, and now identified as the potential organic agricultural production zone, it is necessary to analyse the possible relationship between these two issues, i.e., management of biodiversity and possibility of organic agriculture. So far, most of our initiatives for this region have focussed separately either on biodiversity or on organic farming, rather missing the inherent link between these two. Perhaps the northeast region of India has the unique opportunity to conserve vulnerable plant genetic resources and simultaneously have the strong potential to become the leading centre for organic food production. Therefore, to achieve the goal of agrobiodiversity management which is the most important component of natural resources in northeastern region, through organic agriculture (as an option), several multidisciplinary researches are to be carried out to formulate the proper policy framework. In-depth analysis is required for ethnobotanical and socioeconomic studies to understand the farmers' knowledge. Selection criteria of germplasm and their innovative management are important to know. Such indigeneous knowledge also warrants due share and benefit when it is applicable to advance technology. Dynamics of the local landraces (traditional varieties) and farmers' varieties needs to be understood in detail and, simultaneously, appropriate scientific know-how should be infused with the traditional agricultural practices for optimum exploitation of natural resources and ensuring the synergistic effect of biodiversity-organic agriculture in northeast region.

Horticulture in the Northeast India: The Growth Driver

The northeast region of India comprising the states of Arunachal Pradesh, Assam, Manipur, Meghalaya, Mizoram, Nagaland, Sikkim and Tripura, representing a distinct agroclimatic zone of the country suitable to grow diverse agricultural crops throughout the year. Agriculture and allied sector is the major source of livelihood for the majority of the workforce (around 50-70 per cent across the states). However, agriculture has been remaining, typically, complex, diverse and risk-prone resulting low productivity and income since decades. Under this low-investment—low-income farming situation, the horticulture sector has desirable attributes to accelerate the agricultural growth process in the region. Horticultural sector plays a very important role towards sustainable rural livelihoods in all farming systems in general and in rainfed and hilly farming system in particular like the northeast region. Horticulture today, is not merely a means of diversification but forms an integral part of food and nutritional security (NAAS, 2001). Planning Commission has given special focus to this sector for its holistic development and considerable investment has been made during the Tenth Five-Year Plan (2002 to 2007). Primary objective of this plan investment is to increase area, production and productivity of various horticultural crops in all the northeastern states in accordance with the respective states' underlying potential and in turn to increase the livelihood of the resource-poor farmers. At all-India level, the growing investment in horticulture sector has already started paying dividends through increase in export value of horticultural commodities manyfold and thus, similar is the expectation from the northeast region also. In fact, the signature of the horticultural development in the region has already been seen particularly in the states of Nagaland, Mizoram, Sikkim and Meghalaya. However, the expansion of area and production of horticultural crops has been observed to be restricted in some pockets, where the government intervention is continuing and therefore, the increase in production level is far from the actual potential of the region.

Area and Production of Horticultural Crops

Horticultural crops include production of fruits, vegetables, spices and condiments mainly. Like the trend in all-India level, the progress of area under horticultural crops in the northeast region has been quite encouraging during the preceding decades. In fact, the per cent increase in total area under horticultural crops in the region has been recorded to be around 40 per cent during the period of 1990-91 to 1999-2000 as compared to around 20 per cent at all-India level (Table 7.2). Similarly, during the rest of the period (during 1999-2000 to 2006-07) the increase in area has been even more spectacular (74%) as compared to all-India level (24%). The northeast region is gradually gaining its share in total horticultural area of India which was around 3 per cent during 1990-91 and presently accounted for more than 5 per cent (2006-07). Overall the area has increased from 432 thousand ha to 1,045 thousand ha during the period of 1990-91 to 2006-07. The increase in area under horticultural crops has markedly increased during the year 2003-04 onwards. The reason may be due to the persistent investment in the horticultural sector through technology mission. Almost all the states of the region have shown significant increase in areas under horticultural crops except Tripura and Nagaland during the recent period (during 1999-2000 to 2006-07). Sikkim, Manipur, Mizoram and Arunachal Pradesh have recorded more than 100 per cent increase in area under horticultural crops. Similar to area under horticultural crops, the total production of these crops has also increased significantly during the recent years from around 5,604 thousand tonnes

(2003-04) to 8,847 thousand tonnes (2006-07). Currently the region is accounted for production of nearly 5 per cent of total horticultural produce in India. All northeastern states of the region has shown steady increase in production of horticultural crops. Assam is the major producer of horticultural crops (production of 6,241 thousand tonnes during 2006-07) followed by Tripura, Meghalaya, Manipur, Mizoram, Arunachal Pradesh, Nagaland and Sikkim. The magnitude of increase in production of horticultural crops in the region has been quite encouraging and even has surpassed the increase at national level during recent period indicating the untapped potential of the region. Besides looking into the overall area and production performance of the horticultural crops, it is also important to analyse the performance of various sub-sectors namely, fruits, vegetables, spices and condiments in the region in detail.

Table 7.2

Area and Production of Horticultural Crops in the Northeastern States of India

States	*Area ('000 ha)*					*Production ('000 tonnes)*	
	1990-91	*1999-2000*	*2006-07*	*% Change (1990/91 to 1999/2000)*	*% Change (1999/2000 to 2006/07)*	*2006-07*	*Average Production during 2003-04 to 2006-07*
Arunachal Pradesh	33	38	86.1	15.15	126.58	224.3	221
Assam	209	313	612.4	49.76	95.65	6241.3	4728
Manipur	24	19	49	-20.83	157.89	264.5	341
Meghalaya	57	68	85.7	19.30	26.03	630.9	593
Mizoram	9	20	43.2	122.22	116.00	215.1	139
Nagaland	9	32	30.6	255.56	-4.38	179.5	173
Sikkim	7	10	61	42.86	510.00	171.7	146
Tripura	84	102	76.7	21.43	-24.80	920.1	905
Northeast region	432	602	1044.7	39.35	73.54	8847.4	7245.7
% to all India	3.38	3.93	5.50	-	-	4.7	4.2
All India	12800	15300	18979.6	19.53	24.05	186872	173381

Source: GoI (2004) and National Horticulture Board (2006).

Performance of Fruits

The northeastern region is well known for production of a number of high value fruits across various states, mainly, banana, citrus, pineapple, papaya, passion fruit, strawberry and a number of underutilised fruits. At regional level, the northeast region is accounted for around 6 per cent of total fruits area (300 thousand ha) of India (5,078 thousand ha) and producing nearly 5 per cent (1,577 thousand ha) of total fruit production of India (28,632 thousand ha) (Table 7.3). Based on the average production during the period of 2003-04 to 2005-06, among the northeastern states Assam is having the maximum area under fruit crops (97 thousand ha) followed by Manipur (51 thousand ha), Arunachal Pradesh (50 thousand ha), Tripura (32 thousand ha) Meghalaya (25 thousand ha) and other states. Considerable increase in area under fruit crops during the period of 1991-92 to 2005-06 has been recorded particularly for the states of Manipur (159%), Arunachal Pradesh (156%), Nagaland (154%) and Mizoram (128%). The overall increase in production of fruits has been calculated to be around 68 per cent during the same period (1991-92 to 2005-06), however, which was less than the all-India production increase (105%). All the northeastern states except Sikkim have recorded substantial increase in fruit production. Manipur, Nagaland and Arunachal Pradesh have recorded substantial increase in production. However, the real concern for the fruit crops in the northeastern states is the low productivity (7.69 tonnes/ha) which is far below than the national average (10 tonnes/ha). Of all the northeastern states Tripura (16 tonnes/ha) and Assam have only recorded higher productivity (13 tonnes/ha) than the national level. Productivity of fruits in Meghalaya (9 tonnes/ha) is close to the national level. In fact, the congenial agroclimatic condition of the region is expected to favour higher productivity. More concern is that the fruit productivity of the north-eastern states has been either declining or remaining stagnant during the preceding decades. The main reasons for the low productivity may be: (i) almost all the fruit crops are cultivated in traditional method and lack commercial mode of production; (ii) lack adequate irrigation facilities and fruit production is purely based under rainfed condition; (iii) inorganic input use is almost nil; and (iv) high degree of presence of diversity in fruit crops impede the productivity enhancement because in general diversity in crop germplasm is negatively correlated with the production improvement.

Table 7.3

Performance of Fruits in the Northeastern States of India

States	Area ('000 ha)			Production ('000 tonnes)			Productivity (tonnes/ha)		
	1991-92	Average during 2003-04 to 2005-06	% Change during 1991-92 to 2005-06	1991-92	Average during 2003-04 to 2005-06	% Change during 1991-92 to 2005-06	1991-92	Average during 2003-04 to 2005-06	% Change during 1991-92 to 2005-06
Arunachal Pradesh	20.2	50.3	155.9	47.3	103.2	122.2	2.3	2.0	-13.0
Assam	73.3	97.3	42.3	886.4	1233.1	54.2	12.3	12.6	6.5
Manipur	19.8	51.2	158.6	43.0	320.9	646.3	2.2	6.3	186.4
Meghalaya	24.2	25.3	16.9	218.1	210.3	6.2	9.0	8.3	-8.9
Mizoram	9.3	21.2	128.0	34.8	42.5	22.1	3.7	2.0	-45.9
Nagaland	5.2	13.2	153.8	9.2	48.9	431.5	1.8	3.7	105.6
Sikkim	7.7	9.0	15.6	18.8	11.2	-29.8	2.4	1.3	-37.5
Tripura	44.9	32.1	-26.5	319.1	505.3	64.6	7.1	15.8	123.9
Northeast region	204.6	299.6 (6)	52.4	1576.7	2475.3 (5)	68.4	7.7	8.3	10.5
All India	2874.5	5078.0	91.7	28632.0	51750.5	105.2	10.0	10.2	7.0

Note: Figures in parentheses indicate per cent to respective total to all India.

Source: GoI (2004) and National Horticulture Board (2006).

Special focus is urgently needed to increase the productivity of fruit crops through supply of quality planting material, capacity building of the farmers and to facilitate adoption of improved fruit production methods.

Performance of Vegetables

The northeastern states are congenial for production of a number of vegetables and the market value of these produce become significantly higher when produced during off-season. Major vegetables grown with substantial area in the region are potato, cabbage, cauliflower, tomato, radish and brinjal. Overall the region accounted for 5 per cent of total cropped area under vegetables and producing nearly 3 per cent of total production in India (Table 7.4). The area under vegetables at regional level has increased by around 12 per cent (from 329 thousand ha to 343 thousand ha) as against 28 per cent at all-India level. However, the production increase has not been quite satisfactory (less than 1%). The progress in vegetable production has not gained momentum as it has been observed for fruit production. Sikkim, Meghalaya and Nagaland have recorded substantial increase in area (132%, 61% and 45%, respectively) and production (77%, 294% and 88%, respectively of vegetables during the period 1991-92 to 2005-06). However, for other states the vegetable production has been either of marginal increase or declining. The major factors affecting the vegetable production is the wide fluctuation in prices in local markets due to supply from neighbouring states like West Bengal and also from within the region, Assam. The vegetable production is mainly determined by the demand in major cities within the region where the population is high. Therefore, the vegetable production is primarily confined to the capital cities of the various northeastern states only. The opportunity of commercial vegetable production in rural areas is mainly impeded by the poor road transportation facilities, which is essential to shift these perishable commodities in quickest possible time. The productivity of vegetable production in the region is substantially low (8.7 tonnes/ha) as compared to all-India level (15 tonnes/ha), which is the major concern for increasing the production level of vegetables in the northeast region. The overall productivity of vegetables across the northeastern states is observed to be either stagnant or declining except for Mizoram and Manipur during the period 1991-92 to 2005-06. The major factors affecting

Table 7.4

Performance of Vegetables in the Northeastern States of India

States	Area ('000 ha)			Production ('000 tonnes)			Productivity (tonnes/ha)		
	1991-92	Average during 2003-04 to 2005-06	% Change during 1991-92 to 2005-06	1991-92	Average during 2003-04 to 2005-06	% Change during 1991-92 to 2005-06	1991-92	Average during 2003-04 to 2005-06	% Change during 1991-92 to 2005-06
Arunachal Pradesh	17.1	20.3	19.3	79.9	79.6	-1.3	4.7	3.9	-17.0
Assam	222.4	207.8	3.5	2132.2	1954.0	-11.7	9.6	9.5	-14.6
Manipur	11.8	13.4	13.6	50.3	86.0	71.0	4.3	6.4	48.8
Meghalaya	25.9	35.7	61.0	219.2	293.7	55.2	8.5	8.3	-3.5
Mizoram	6.0	4.6	-61.7	31.8	26.5	-1.3	5.3	7.3	152.8
Nagaland	8.2	11.9	45.1	66.9	88.1	31.7	8.2	7.4	-9.8
Sikkim	7.6	16.9	131.6	46.1	76.9	71.6	6.1	4.6	-26.2
Tripura	30.3	32.4	7.6	306.8	366.4	21.8	10.1	11.3	13.9
Northeast region	329.3	343.1 (5)	12.4	2933.2	2971.2 (3)	0.9	8.9	8.7	-10.2
All India	5592.7	6739.0	28.1	58532.0	101153.4	86.3	10.5	15.0	44.8

Note: Figures in parentheses indicate per cent to respective total to all India.

Source: GoI (2004) and National Horticulture Board (2006).

low productivity of vegetables in the region is because the production system is constrained with assured irrigation (mainly cultivated as rainfed) facilities, low use of inorganic inputs and pesticides. However, in some pockets (like East Khasi Hills in Meghalaya) the use of inorganic inputs and pesticides has been observed to be very high along with high productivity as these farmers are cultivating vegetables in commercial mode. Therefore, there is a distinct dichotomy in vegetable production in the various pockets of the region and marked difference is there in terms of productivity.

Performance of Spices and Condiments

The northeastern states have distinct advantage in production of a number of high value spices such as ginger, turmeric, large cardamom, chillies, black pepper etc., which have substantial demand in domestic as well as international market. The export potential of these crops are enormous and usually these crops are considered to be the major cash crops for the livelihood of the farmers. Overall the region has around 156 thousand ha of area under spices & condiments and producing more than 400 thousand tonnes of these high value commodities (Table 7.5). At individual state level within the region, Assam accounted for more than 70 thousand tonnes followed by Sikkim, Manipur, Meghalaya and other states. Similarly, Assam is producing nearly 180 thousand tonnes of spices and condiments followed by Meghalaya (60 thousand tonnes), Sikkim (42 thousand tonnes), Arunachal Pradesh (39 thousand tonnes), Mizoram (39 thousand tonnes) and Nagaland (32 thousand tonnes). Very low production is recorded in the database of Spice Board that possess some valid doubt for the total spices production in Manipur. Among the spices & condiments ginger is the most important crop irrespective of states in the northeast region and accounted for more than half of the total area under all spices & condiments, followed by chilli and turmeric. In Sikkim, large cardamom is the most important spice crop and accounted for around 79 per cent of the total area under spices & condiments. The value of these spices and condiments can substantially be increased when produced organically which have large export demand within the country as well as around the developed countries. The initiative and process of organic farming production has already started in this region through active participation of Spices Board, respective state government, various agencies (like

Table 7.5

Area and Production under Spices & Condiments in the Northeastern States (Three Years Moving Average)

States	Area under Total Spices ('000 ha)		Production of Total Spices ('000 tonnes)		% Area of Major Spices within Total Area under Spices (during 2003-04 to 2005-06)		
	During 2000-01 to 2002-03	During 2003-04 to 2005-06	During 2000-01 to 2002-03	During 2003-04 to 2005-06	Chilli	Ginger	Turmeric
Arunachal Pradesh	6.6	6.93	39.38	38.68	26	67	7
Assam	70.75	71.41	168.53	179.33	20	25	17
Manipur	11.51	13.43	8.3	9.93	71	15	2
Meghalaya	19.9	13.52	72.95	60.8	10	49	10
Mizoram	7.76	6.23	39.2	38.88	22	72	4
Nagaland	2.7	5.72	20.88	32.6	44	33	9
Sikkim	30.98	33.85	30.56	41.92	*	19	2
Tripura	-	4.4	-	14.93	38	31	31
Northeast region	152	155.58	379.8	417.00	21	29	11

Note: * In Sikkim 79 per cent of total spice area is accounted by large cardamom.

Source: Spices Board India (2007).

Agricultural & Processed Food Products Export Development Authority (APEDA)) as well as research organisation such as Indian Council of Agricultural Research (ICAR). But the region has to resolve several issues, like promotion of inspection and certification in compliance with the National Programme for Organic Production (NPOP)/destination countries, maintaining quality standard, arranging proper packaging and distribution of these crops etc., to take forward the organic farming movement. Nevertheless, spices and condiments have the highest potential to be grown organically from this region.

Output Growth of Horticultural Crops

Having analysed the performance of horticultural sector in the north-east region through change in area, production and productivity of fruits, vegetables, spices and condiments, it is important to examine the value of output growth of major horticultural commodities across the various states in the northeast region. The value of output data has been taken from the database 'Statewise Estimates of Value of Output from Agriculture and Allied Activities' published by Central Statistical Organisation (CSO), Government of India in various years. The value of output growth for the fruits and vegetables was observed to be substantially higher in all northeastern states (except Arunachal Pradesh and Assam) in comparison to all-India level (5.43%) (Table 7.6). Nagaland has recorded more than 20 per cent annual compound growth rate for total fruits and vegetables production followed by Mizoram (15%), Tripura (9%), Meghalaya (7%), Sikkim (6%) and Manipur (4%) during the period 1990-91 to 2002-03. Within the agriculture sector horticulture sector has been showing substantial growth potential across the northeastern states. Output growth of ginger and other horticultural crops have been observed to be quite encouraging across the various states of the region (see Annexure Table A-7.1). Output growth in potato production has been quite satisfactory (ranging between 3-13 per cent across the states) except for the states of Sikkim, Manipur and Arunachal Pradesh. It is quite interesting to observe substantial growth in value of output (10–32 per cent across the states except Sikkim) for the horticultural commodities produced in kitchen garden. In northeastern region the horticultural crops are produced through farming system approach without any commercial mode of production.

Table 7.6

Value of Horticultural Crops in Northeastern States of India

States	Value in Rs. Lakh at Current Prices (2005-06)			Annual Compound Growth Rate (%) during 1990-91 to 2002-03				
	Total Spices & Condiments	Other Horticultural Crops	Total Fruits & Vegetables	Total Agriculture	Total Spices & Condiments	Other Horticultural Crops	Total Fruits & Vegetables	Total Agriculture
Arunachal Pradesh	6045	15022	18216	52444	13.51	-12.43	-11.13	-4.32
Assam	135979	219557	327384	1076426	3.23	4.84	3.69	2.63
Manipur	3302	28323	37480	86929	5.45	4.56	4.16	2.46
Meghalaya	9982	31516	53464	90200	3.32	10.57	6.67	2.22
Mizoram	4891	6202	7458	33050	14.55	29.04	15.26	4.72
Nagaland	11883	6484	16065	86631	23.56	19.92	20.33	12.57
Sikkim	9646	5627	29721	50584	-0.25	12.18	5.55	5.22
Tripura	7322	81063	94364	184822	8.53	9.69	9.29	0.31
All India	1673592	9819121	13367064	51881281	4.35	5.63	5.43	2.02

Source: Estimated from CSO (2006) *Database on State-wise Estimate of Value of Output from Agriculture and Livestock* (at 1993-94 constant prices).

Farmers try to maximise their returns from the farming system through incorporation of as many as of horticultural crops and ultimately allocation of areas for individual crops become very small. Similarly, since the region is full of agrobiodiversity large number of horticultural crops are being produced by the farmers with low volume which are considered to be underutilised horticultural crops and included in the 'other horticultural crops'. Because of their large number these 'other horticultural crops' have substantial contribution in total horticultural production. Selection of potential (of mass production) underutilised crops (e.g., such as local pear or *kompe*, sohiong, puri, etc.) needs special attention to increase the overall production of the horticultural crops in the region.

Investment and Achievement in Horticulture Sector

Viewing the growth potential of the horticultural sector, Planning Commission has been continuing high focus through increasing the investment in many folds. At all-India level the investment in horticulture sector has been increased to Rs. 789 crore during Eighth Five-Year Plan (1992-1997) from Rs. 24 crore during Seventh Plan (1985-1990) (Table 7.7). The investment in horticulture sector has been further increased to Rs. 1,453 crore during Ninth Plan (1997-2002) and in continuing with the thrust in this sector the investment during Tenth Plan (2002-2007) has been made as high as Rs. 4,472 crore. Special budget has been allocated to develop the horticultural sector in the northeastern region through implementation of technology mission during Tenth Plan. Through this technology mission, so far Rs. 858 crore has been spent in all the northeastern states during the Tenth Plan period. Therefore, it can be stated that like in all-India level, the northeastern region also has received special attention to develop the horticulture sector and considerable amount of investment has been made. As against this investment, it is important to look into the achievement made through the implementation of the technology mission in the region (Table 7.8). As per the cumulative progress report given by the individual states, the achievement through implementation of the technology mission has been substantial in terms of area expansion under major horticultural crops such as, fruits, vegetables, spices, plantations, aromatic plants and flowers. Overall the area expansion under fruit crops has been calculated to be over 50,000 ha for the region followed by over 24,000 ha under spices, over 22,000 ha under vegetables, over 4,000 ha under plantation crops and more than 2,000 ha under medicinal crops as well as aromatic crops both. The area expansion through the technology mission on horticultural crops has been observed to be encouraging in the NER. In case of fruits, area expansion has been more in Mizoram, Arunachal Pradesh and Meghalaya. Similarly, for vegetables Tripura, Sikkim and Arunachal Pradesh progress well and for spices performance in terms of area expansion was high in Sikkim and Arunachal Pradesh. Other states also have recorded substantial increase in area under horticultural crops through this technology mission. However, one important note of caution is that so far this area expansion has been done through continuous investment and government efforts, therefore, the long run implication of this investment, particularly the sustainability of these activities has to be analysed carefully to understand the real impact of this technology mission once the flow of fund is being stopped. One of the reasons for lack of enthusiasm for the investment in the region has been observed that in general the autonomous investment under various project activities had failed to attract induced investment due to lack of private initiative and therefore, once the project is being withdrawn the fate of the project also becomes bleak.

Table 7.7

Plan Investment for Horticultural Development in India vis-à-vis Northeast Region

Five-Year Plan (FYP)	*Investment (Rs. Crore)*	*Investment through Technology Mission (Rs. Crore) in the Northeast*		
		Budget Exp.	*Revised Exp.*	*Expenditure*
Fifth FYP	7.61	*	*	*
Sixth FYP	9.13	*	*	*
Seventh FYP	24.19	*	*	*
Eighth FYP	789	*	*	*
Ninth FYP	1453	*	*	*
Tenth FYP	4472.37	829.62	890.41 (P)	858.37 (P)
2002-03	283.15	120	91	91
2003-04	291.22	134.22	149.51	118.39
2004-05	542	200	171.5	176.72
2005-06	1405	170	185 (P)	187.54 (P)
2006-07	1951	205.4	293.4	284.72

Note: * Data was not available separately for the northeast region.

Source: Government of India (2008).

Table 7.8

Achievement of Technology Mission in Horticultural Development in Northeast Region of India (2001-02 to 2005-06)

States	*Area Expansion under Horticultural Crops (ha)*						
	Fruits	*Vegetables*	*Spices*	*Plantation*	*Medicinal*	*Aromatic*	*Flowers*
Arunachal Pradesh	10987	2460	5570	0	1000	1317	1140
Assam	3694.3	949.4	2079.9	1240	117.5	92.5	404.6
Manipur	4500	1298	833	55	72.5	112.5	35
Meghalaya	6027	2125	1966	750	196.5	116.5	783
Mizoram	11700	3177	2111	670	215	234	5338
Nagaland	4024	1280	2540	1245	428	574	529
Sikkim	4166.7	4533	7613	0	35	28	2089
Tripura	5827	6903	1751.4	719.6	18	9	347.4
Northeast region	50926	22725.4	24464.3	4679.6	2082.5	2483.5	10666

Source: Government of India (2008).

Way Ahead

The investment in horticultural sector in the northeastern states has started paying dividend but at a slower rate in comparison with its underlying potential. Development of fruit sector has been observed to be encouraging but same is not true for the growth in vegetable production. And the spices production have been growing at a slow rate or remaining almost unchanged across the various states of the region. Mass production and distribution of quality planting material for fruits and spices can be given top priority. Special support and focus needs to be given to capacity building of the farmers to produce vegetables under protected as well as open condition. Expansion of area under off-season vegetable crops should be made around the various hilly tracts of the region. Application of nutrient and water management may be given special focus to improve the productivity of horticultural crops particularly vegetables. Farmers/growers association, contract farming (particularly for organic production) and production through self-help groups (SHGs) may be promoted to increase the volume of production, which in turn would facilitate the marketing of these high value crops. One of the major problem in growth of horticultural sector in the region is the lack of sustainability of any projects—once the government intervention is withdrawn, the farmers either disengage or fail to continue their activities primarily due to their resource-poor condition. Providing loan at a special rate of interest (may be four per cent) can be attempted to address this issue. Appropriate methodologies are needed to estimate the area, production and productivity of horticultural data and simultaneously, creation of detailed database is strongly advocated. Finally, establishment of market intelligence unit is utmost important to provide advisory services of price movement and market arrival of various horticultural commodities through recording and analysing the day-to-day, weekly, monthly, seasonally or yearly data.

Trade Opportunities for Livestock Sector in the Northeast

Livestock sector has the desired growth prospect to accelerate and sustain the agricultural development in the northeast region. Among the livestock resources, pig is one of the important animals within the livestock sector of northeastern states and play major role in the livelihood of the tribal people. However, these are mostly reared as an integral part of their farming system without any commercial mode of production failing to trigger the desired level of dynamism to capitalise the real potential. Mostly the states have the deficit production of pig in view of high local preference of the meat and leaving almost no surplus for external trade. In spite of having the growth potential there are several factors of cost disadvantage that accrue to domestic producer in these states such as infrastructure, power, internal freight, inverted duty structure which needs to be addressed appropriately particularly under the context of liberalised market economy. Moreover, the people of northeast are considered to be underprivileged and marginalised since ages and it may be logical to apprehend that the rapid marketisation of Indian economy may aggravate the situation adversely unless appropriate safeguard measures are taken care of. Therefore, careful analysis is essential to identify the possible threat/opportunity and/or gain/loss particularly in view of the opening up of Indian market. The present chapter tries to highlight these issues by focussing on the trade of pig meat with a special reference to the northeastern states of India.

Contribution of Meat in Agriculture Sector

Meat in the form of beef, mutton, pork or poultry contributes substantially (over 15%) to the total value from the livestock sector and in turn to the agriculture and allied sector as a whole (nearly 4%). Poultry meat continued to be the highest contributor to the total meat production in India followed by mutton, beef and pork. Pork contribution to total value of meat production is observed to be hovering around 5-6 per cent. The contribution of pork in total meat production may appear to be meagre at national level, but has significant contribution when compared at regional level particularly northeastern region. Production estimates from Integrated Sample Survey Scheme for Estimation of Major Livestock Products (Ministry of Agriculture) indicated that pork contribution to total meat in Mizoram is as high as 70 per cent and in other northeastern states the share of pork to total meat ranged between 27-53 per cent. Therefore, it is explicit from the above discussion that at national level pork has smaller contribution to total meat but at northeast region level it has substantial contribution to total meat production and plays important role in agricultural production system.

Pig Population around the Indian States

At national level pig population accounted for nearly 3 per cent but at northeast region level it has the larger share to the tune of 28 per cent indicating the importance in total livestock resources. Along with the northeastern states, other states, namely, Uttar Pradesh, West Bengal, Jharkhand, Bihar and Orissa have a larger pig population base and also exhibits positive change in pig population over the period of time, except Uttar Pradesh. It is quite notable from the distribution of pig population around various states that its share is high and changing positively in the states where the population base for downtrodden societies (SC and ST) is also high. This implied that pig is mainly distributed in the areas or pockets with high concentration of underprivileged people and therefore, needs to be addressed with special consideration particularly in view of the economic liberalisation to safeguard their livelihoods. It is also notable that the poverty ratio is quite high in these states and the production system of the people belonging to the weaker section may not be competent enough to take advantage of economic liberalisation. This emphasise the need for careful policy prescription to safeguard their livelihood that otherwise would have an adverse effect which invariably will increase the social tension. Since pig is the important constituent of the total livestock resources of these underprivileged people, linking the farmers to the trade (national as well as international) through suitable policy initiatives (such as contract farming, PPP, future trading etc.) may articulate the desired vibrancy and dynamism to this sector which in turn can contribute positively towards the welfare of these farmers. In view of this it is essential to understand the trend in meat consumption in domestic sector as well as possible opportunities in other countries also.

Domestic Consumption of Meat

Bovine meat was accounted for the nearly half of the total meat consumption in domestic market followed by sheep/goat meat, pork and chicken during 1990s. But the chicken meat consumption increased by nearly 12 per cent during 1990-2004 and presently accounted for 29 per cent of total meat consumption, thereby, pushing down the share of pork to 9 per cent. The trend in domestic meat consumption is highly skewed, bovine meat and the share of chicken meat is also increasing rapidly. The share of pork in total meat consumption is almost hovering around 10 per cent. The possible reason might be that either the supply of pork is increasing at a slow rate or the consumer base for pork is not changing significantly in domestic market. This implied that possibly the pork is produced and consumed by certain group of people and thereby, demand and supply situation is also almost static. This emphasise the need for expansion of domestic production as well as establishing the linkages with the external market, where the consumers have high preference for pork, such as neighbouring countries of northeast region. In addition there is need for establishment of processing centre for creation of value addition to the pork so that it can be readily acceptable to the final consumer.

Opportunities for Pork Export

India's share in external trade of pork is very small and negligible as compared to other Asian countries such as China, Thailand, Vietnam, the Philippines and Japan. However, of late India's exports is increasing at a slow rate and the trade balance is positive as far as pork is concerned.

There is ample opportunity to increase India's pork exports particularly to the countries like China and Japan through pushing the supply forces through policy intervention. China's imports is increasing at a faster rate than exports whereas Japan is the major importer of pork among the Asian countries. However, to increase the pork exports to this country it is extremely important to follow the sanitary and phytosanitary measures (SPS) in compliance to that country for which the production of pork should be made more organised and calls for investment in processing and packaging. The technical barrier to trade (TBT) and non-trade barrier would play the most crucial role for external trading of pork. This problem is more pronounced in the backward regions like northeastern states where the production system is still operating mostly at subsistence level. In spite of having these bottlenecks still India has the comparative advantage of producing pork as far as producer price is concerned. In fact, the production cost of pork in India is more or less 3-10 times cheaper than other Asian countries. Also it is encouraging that the cost of production of per tonne of pork in India is declining by nearly three per cent during the period of 1991-2003. This implies that even though the pork production system in India is under unorganised sector and not having competitive advantage, however, posses the advantage of low cost of production and perhaps can deliver positively when pushed through desired policy initiatives. But the real concern is—in spite of having the low cost of production advantage; Indian pork producer has several factors of cost disadvantage such as poor infrastructure, irregular power supply, internal freight and high transportation cost which accrue to the domestic producer. While framing the policies for liberalising the meat industry, it is essential to analyse these cost disadvantages carefully to safeguard the interest of the domestic producer. It is extremely important to examine the contribution of pig to the livelihood security of the domestic producer to regulate the pork industry of the country. Besides, market research reports reveal high market growth for the organic meat sector as consumers' fear for food safety raise demand for organic meats. For example organic meat market in Canada is expected to grow by 35 per cent. Similar growth was expected for US market from 2004 as organic meat production steps up and as more volume goes into the retail trade. Organic poultry was the most widely available organic meat in 2003 with organic beef and pork rarely found in retail. Currently organic meat is less than one per cent of the overall US

$175 billion of US beef market. However, interest in organic meat is high among the consumers in US with the Organic Trade Association, in Greenfield, Massachusetts, that projected a 175 per cent growth rate for organic meat by 2005 (EXIM Bank of India, 2006). However, along with the high demand for organic meat there is also a steep competition among the suppliers. Therefore, success of organic meat movement will depend how best the Indian meat industry can be linked with the international market, equipped with continuous flow of market intelligence and finally how fast the production pattern can be internalised along with the changing demand for the products.

Constraints in Production and Trade

Despite of having resources in terms of live animals and potential the pork supply in northeastern region failed to grow at a desired pace due to several factors which includes both supply-side as well as demand-side constraints. In fact it is quite contrasting that even though India has positive trade balance in pig meat export-import but simultaneously, has negative trade balance (deficit of US $0.125 million during 2002 as far as trading of live pig is concerned (FAO, 2005). The most important constraint can be attributed to the non-availability of feed for pig at adequate quantity and at reasonable prices. The price of concentrated feed which can increase the body weight substantially is neither available nor affordable to the resource-poor farmers in these states. The pigs are mostly reared as an integral part of their farming system without any special care purchased feed. Though green fodder is abundant in these states, however, dry fodder is either not available or costly for the farmers. Another important constraint is the low share of crossbred pig in total pig population in these states. In most of the states the pig belongs to the indigeneous group and thereby, annual average yield is also very low. Also the high mortality rate at farm level causes the great loss to the farmers as swine fever is quite prevalent in these states. The demand-side constraints (preference of pork by definite group only) also contribute negatively towards the production of pig for domestic market.

The pig is an integral part of the farming system of the farmers particularly for the weaker sections of the society. Pork has the substantial potential to grow and contribute positively to increase the trade balance of meat which in turn may be helpful to increase the wellbeing of the underprivileged farmers. However, this sector needs appropriate policy attention both for supply side as well for demand side. While formulating the policies for the pig sector, it is essential to know the possible opportunity and threat to the livelihood of the resource-poor farmers for external trading. It is important to know the magnitude of the cost disadvantage to safeguard the greater interest of these farmers. Studies are essential to know whether there is a need for special protection for pig and its products. Consumer preference to the various forms of meat indicated that pork is produced and consumed by certain group of people and thereby, has limitation to grow at a faster rate as compared to other meat as far as domestic consumption is concerned. However, the opportunity of external trade with the other Asian countries can be further explored and capitalised.

Annexure Table A-7.1

Value of Output Growth of Potato, Dry Ginger and Kitchen Garden Crops in NER vis-à-vis India

States	Output Growth in % (during 1990-91 to 2002-03)		
	Dry Ginger	Potato	Kitchen Garden
Arunachal Pradesh	18.34	-0.38	12.23
Assam	7.12	3.91	10.51
Manipur	4.40	-0.95	9.89
Meghalaya	2.69	2.61	22.75
Mizoram	17.27	9.97	31.59
Nagaland	21.71	13.31	3.22
Sikkim	-5.33	-3.93	-2.15
Tripura	8.34	5.63	14.37
NER	7.41	3.53	8.90
India	5.50	5.17	5.61

Source: Estimated from CSO (2006) *Database on State-wise Estimate of Value of Output from Agriculture and Livestock* (at 1993-94 constant prices).

8 Opportunities of Production of High Value Crops through SHGs

A Case Study of Meghalaya

Diverse agroclimatic condition of Meghalaya is highly suitable for growing a number of high value horticultural crops. However, low volume of marketable surplus and lack of assured supply of these high value products throughout the year, constrained the development of desired market dynamics in the state. Horizontal integration among the farmers/growers has been viewed as one of the options to off-set these shortcomings of present marketing system. Through self-help groups (SHGs), farmers are encouraged to grow crops collectively so that the volume of production is increased and thereby facilitate marketing of agroproducts through active participation of various agencies. Key institutions and organisations such as State government, National Bank for Agriculture and Rural Development (NABARD) and non-governmental organisations (NGOs) are taking active participation to facilitate the functioning of these informal groups at grassroot level. So far the total of 4,843 SHGs have been formed in the state covering 5,2910 members. Out of which 42 per cent SHGs are run by female members only. In Ri-Bhoi district, so far 426 SHGs have been formed with a total membership of 4,467 among which nearly half of the SHGs are operated by female members. The present chapter tries to analyse the functioning of such SHGs in Ri-Bhoi district of Meghalaya focussing on some of the important high value crops namely, ginger, capsicum, tomato, turmeric, broccoli, strawberry, anthurium, rose, lilium, etc.

This case study is based on primary data colleted from different farmer groups growing these crops at four villages (Mawkyrdep, Umranong, Umtham and Nongkya) in Ri-Bhoi district of Meghalaya. The information was collected based on pre-structured tested farm survey schedule from 40 farmers engaged in production of these crops during 2005 to 2007.

Current Status and Distribution of SHGs in Ri-Bhoi District

Ri-bhoi district comprises three blocks, namely, Jirang, Umling and Umsning. So far a total of 426 SHGs have been established in Ri-Bhoi district covering 4,467 members (Table 8.1). Nearly 10 per cent of total SHGs in Meghalaya are functioning in Ri-Bhoi district. Nearly 50 per cent of the SHGs are operated by female members only followed by male (38 per cent) and rest are operated by mixed type of membership (male and female both). Most of these SHGs have been established during 2001-2005. Formation of more than 80 per cent of these SHGs has been initiated by the well-known NGO, Bosco Reach Out (BRO). Rest of the SHGs have been formed by the initiative of various governmental organisations such as Block Development Office (BDO), District Rural Development Agency (DRDA), Department of Soil and Water Conservation and Department of Social Welfare. These SHGs are engaged in diverse income generating activities of agriculture which includes production of turmeric, banana, pineapple, orange, ginger, areca nut, betelvine, strawberry, vegetables, potato, black pepper etc. Besides, many SHGs are engaged in production of livestock and other activities (e.g., weaving and handloom, production of medicinal and aromatic plants, mushroom, flower making etc.). Within agriculture, some high value crops like production of ginger and vegetables are very popular among the farmers and plays significant role in the farm households economy. Therefore, detail studies have been carried out to know the economics and scale of operation of these activities.

Table 8.1

Distribution of Self-Help Groups (SHGs) in Ri-Bhoi District (Block-wise) of Meghalaya

Particulars	*Jirang*	*Umling*	*Umsning*	*District Total*	*State Total*
Total no. of SHGs	112	128	186	426	4843
No. of SHGs operated by male	66	30	68	164	844
No. of SHGs operated by female	38	90	80	208	2010
Mixed (male plus female)	8	8	38	54	949
Total membership	1249	1307	1911	4467	52910

Source: Primary Survey (2007-08), Division of Agricultural economics and Statistics, ICAR Research Complex for NEH Region, Umiam (Meghalaya).

Production and Marketing of Ginger by Sample Farmers

Ginger is one of the most preferred cash crops in Meghalaya and all sampled farmers (40) were observed to be growing ginger at various scales (Table 8.2). Quantity of ginger production ranges from less than 100 kg and up to 2,000 kg. Nearly half of the sample farmers are producing ginger in the range of 250-500 kg. On an average individual farmers are producing 828 kg of ginger. Out of which nearly 75 per cent (622 kg) of their produce is being marketed. Average volume of ginger marketed ranges from as low as 45 kg up to 1,600 kg. Marketed surplus has been calculated to be varying from 70-83 per cent across various production ranges. Average price received by the farmers were observed to be Rs. 9 kg.

Table 8.2

Average Volume of Production and Marketing of Ginger by Sample Farmers

Srl. No.	*Production Range (kg)*	*No. of Farmers*	*Average Volume of Production (kg)*	*Average Volume of Produce Marketed (kg)*	*% of Produce Marketed*	*Average Price Received (Rs./kg)*
1	< 100	4	65	45	69	8
2	100 – 250	6	207	167	81	9
3	250 – 500	18	389	322	83	8
4	500 – 1000	2	1000	800	80	9
5	1000 – 1500	6	1307	900	69	9
6	1500 – 2000	4	2000	1600	80	9
7	Overall	40	828	622	75	9

Note: Total production of ginger by the sampled farmers (40) was calculated to be 331.20 q.

Source: Primary Survey (2007-08), Division of Agricultural economics and Statistics, ICAR Research Complex for NEH Region, Umiam (Meghalaya).

Cost and Returns from Ginger Cultivation

Estimation of annual cost and returns of ginger in Meghalaya is quite difficult due to the problem of estimation of area. In view of this problem, the per hectare cost and return of ginger cultivation have been estimated by pooling all cost items as well as adding volume of production together. Total cost and gross return was calculated on the basis of per ha production of ginger. Ginger was observed to be cultivated by all the farmers in the sampled group with varying quantities ranging between less than 100 kg up

Table 8.3

Costs and Returns of Ginger Cultivation on Sampled Farms in Meghalaya

(Rs./ha)

Srl. No.	*Particulars*	*Quantity*			*Costs and Returns (Rs.)*	*Share to Total (%)*
		Male Labourers	*Female Labourers*	*Total Labourers*		
1	Land preparation and tillage (man-days)	58	45	103	8950	22.76
2	Sowing & transplanting (man-days)	20	14	34	2980	7.58
3	Weeding (man-days)	20	25	45	3750	9.54
4	Harvesting (man-days)	24	18	42	3660	9.31
	Total labour	122	102	224	19340	49.19
6	Seed used (kg)		-	1998	19980	50.81
7	Total cost (Rs.)	-	-	-	39320	100
	Yield (kg.)	-	-	8992	-	-
8	Gross returns (Rs.)	-	-	-	80928	-
	Net returns (Rs.)	-	-		41608	-
9	Cost per kg (Rs.)	-	-	-	4.37	-

*Note: Wage rate of male (M) and female (F) labourers are Rs.100 and Rs. 70 per man-days, seed cost @ Rs. 10/kg, selling price of ginger @ Rs. 9/kg.

Source: Primary Survey (2007-08), Division of Agricultural economics and Statistics, ICAR Research Complex for NEH Region, Umiam (Meghalaya).

to 2,000 kg. Together the estimated amount of ginger produced by all the farmers was calculated to be 331.20 ql. and on an average farmers were producing 828 kg of ginger. Total cost of ginger cultivation has been estimated to be Rs. 39,320/ha and the cost per kg was calculated to be Rs. 4.37 (Table 8.3). Major cost components for ginger cultivation include cost of seeds (51 per cent of total cost) and cost of labour (49 per cent). Ginger, being largely grown under *bund* system (raised bed) as rainfed crop in sloppy land by the majority of the farmers without using any external inputs such as inorganic fertiliser and plant protection measures (Tripathi *et al.*, 2006). A total of 224 man-days/ha of labour were employed for the production of ginger with the participation of almost 50 per cent of male and female labour each. Major activities of ginger cultivation are land preparation and tillage, sowing and transplantation, weeding and harvesting. Most of these activities are shared by both male and female workers. Prevailing wage rate was observed to be of Rs. 100 and Rs. 70 for male and female labourers, respectively. Total value of the ginger produced was estimated to be Rs. 80,928/ha and net returns were calculated to be Rs. 41,608/ha. Ginger is considered to be one of the highly preferred cash crops in Meghalaya because of its favourable economics. However, the price fluctuation in different years is a major concern for the farmers. Sometimes the farmers are forced to sell their produce even at the price of less than Rs. 5/kg. Farmers stated that usually this happens once in every three years in Meghalaya. Price of ginger in the state is determined by the volume of production during previous year, supply of ginger from other areas of the country and also export demand situation in the current year.

Production and Marketing of Vegetables

Farmers in the selected villages were observed to be growing a number of vegetables namely, tomato, sweet potato, French bean, peas, chilli, cucumber, cabbage and broccoli. Some of the vegetables are grown after receiving the extension support from the state government particularly District Horticulture Office (DHO). Production of broccoli is being promoted by the DHO through supplying seeds and also ensuring the buy-back of the produce from the farmers. Out of the total (40) sampled farmers six farmers are producing broccoli with an average volume of production of 125 kg (Table 8.4). Marketed surplus for broccoli has been estimated to be 88 per cent (110 kg). The average selling price for broccoli has been observed to be Rs. 19/kg. Farmers usually sell their produce (broccoli) either in local market or to DHO. Cabbage is also grown by the farmers and average volume of production has been estimated to be 150 kg. Out of which 140 kg (93 per cent) is being sold to the market with an average price of Rs. 10/kg. Farmers prefer to sell cabbage in the local market mainly. For the village Mawkyrdep the local market is either Nongthymmai (Kyrdem) or Bhoirymbong which are located within 1-3 km from the village. For other villages (Umranong, Umtham and Nongkya) the local market is Umsning which is located around 10 km from these villages. Cucumber is another vegetable observed to be produced by the farmers in the sampled villages but the volume of production was very less (23 kg only). The average price received for cucumber has been observed to be Rs. 9/kg and farmers mostly sell it at local market. Tomato is a highly preferred vegetable among the sample farmers and the average volume of production has been grouped into two categories i.e., up to 100 kg and more than 100 kg. Since the range of production is very wide, therefore, the farmers have been grouped into two production groups to avoid the misleading average value of production. Within the production range of up to 100 kg the average volume of production was calculated to be 83 kg and similarly, average volume of production for other group was about 172 kg. Under both the categories farmers were observed to be producing entire volume of tomato for selling in the market only. Average price received by the farmers were observed to be Rs. 10/kg and mostly farmers used to sell their produce at Shillong market which is about 30 km away from these villages. French bean is also another preferred vegetable grown by the farmers. Nearly 10 farmers were observed to be growing french bean with an average price of Rs. 12 per kg.

Sweet potato is also another important vegetable grown by the farmers of Meghalaya. Mostly it is used either as vegetable or as feed for pigs. Sweet potato is grown as intercropped with other crops, mainly with ginger. Like tomato production, the range of sweet potato has been categorised into two groups, i.e., production up to 100 kg and production more than 100 kg. Out of the total sampled farmers, eight farmers were observed to be growing sweet potato at the average production of 68 kg and two farmers were producing on an average 250 kg each. Marketable surplus was calculated to be 66 per cent for the first group and the same was nearly 80 per cent for other group of producers. Lower marketable surplus under less than 100 kg production group can be attributed to the need of meeting home consumption. Average price received for selling of sweet potato was observed to be Rs. 6/kg and mostly these are disposed of at local market only.

Chilli was observed to be another vegetable preferred by the farmers to grow mainly due to its high prices. However, the average volume of production was observed to be quite low (only 30 kg) and out of which nearly 90 per cent of the produce was calculated to be as marketed surplus. Average price received was found to be usually high for this year at Rs. 37/kg. Farmers were observed to be selling chilli both at local market as well as in Shillong market. Very high price of chilli and other vegetables during survey period in the current year can be attributed to the short supply of vegetable from neighbouring state due to severe flood in Assam. Meghalaya is heavily dependent on the supply of vegetables from Assam. Peas grown by farmers with average production of 20 kg only and the entire amount of peas was observed to be produced for selling at market mainly. Mostly peas are sold at Shillong market and average price received by the farmers were calculated to be of Rs. 15/kg.

Table 8.4

Production and Marketing of Vegetables of Selected Farmers in Meghalaya

Srl. No.	*Crops*	*No. of Farmers*	*Average Volume of Production (kg)*	*Average Volume of Product Marketed (kg)*	*% of Production Marketed*	*Average Price Received (Rs./kg)*	*Remarks*
1	Broccoli	6	125	110	88	19	Supported by DHO
2	Cabbage	5	150	140	93	10	Sold at LM
3	Cucumber	8	23	20	87	9	Sold at LM
4	Tomato (up to 100 kg)	4	83	83	100	10	Sold at SM
5	Tomato (more than 100 kg)	6	172	172	100	11	Sold at SM
6	French bean	10	52	47	90	12	Sold at SM
7	Peas	2	20	20	100	15	Sold at SM
8	Chilli	6	30	27	90	37	Sold at LM & SM
10	Sweet potato (up to 100 kg)	8	68	45	66	6	LM
11	Sweet potato (more than 100 kg)	2	250	200	80	6	LM

Note: LM and SM stand for local market and Shillong market, respectively. Average distance of Shillong market is 30 km from the villages.

Source: Primary Survey (2007-08), Division of Agricultural economics and Statistics, ICAR Research Complex for NEH Region, Umiam (Meghalaya).

Advantage of Production of High Value Crops through SHGs

Detailed discussion was held with the farmers to know whether they are realising any advantage in production and marketing of these crops through formation of SHGs over the farmers being non-members of any groups. It was felt that there was no significant difference in price realisation of the produce when producing through SHGs or non-SHGs.

This is mainly due to the reason that the SHG movement in the state is still in the nascent stage and the volume of production is also quite low failing to increase their expected bargaining power of the produce. Once the volume of production will increase and the farmers are likely to sell their produce at distant market, which is likely to fetch higher returns. Also the formation of SHGs has increased the investment capacity of the farmers and they are now ready to adopt other high value crops as per the changing market demand situation. In longer period these SHGs are likely to facilitate the institutional borrowing also to meet their additional financial requirement. In addition, the SHGs have made the farmers more confident in growing diversified crops and the risk perception of new technologies has also changed. In nut shell, it was realised while discussing with the farmers that the seeds of hope have been sown through formation of these SHGs and it may take some more time to translate these expectation into benefits.

Constraints in Production of High Value Crops

Detailed discussion was held with the farmers to identify various constraints in production and marketing of high value crops in the study area. Among the production constraints, disease and pest attack particularly for ginger was severe. Few farmers also pointed out that the productivity of ginger also was observed to be declining over a period of 8-10 years. Previously farmers used to get nearly 5-6 times of ginger over the amount of seed used but recently the harvesting was not more than 3-4 times over the amount of seed used. Possible reason might be—under usual practice farmers used to cultivate ginger in every three years and during this period (after harvesting of ginger in the first year), either land is kept fallow or rice followed by vegetables are cultivated in the same land. Farmers follow traditional method of cultivation practices and no external inputs are applied and also no soil nutrient management is followed except natural regeneration of soil fertility. In addition, ginger is a soil exhaustive crop, therefore, under traditional cultivation practices the productivity of ginger observed to be declining. Deep virgin soils of forest brought under *jhum* system are giving higher yields in first and second year of cultivation even under zero nutrient management conditions. But heavy rains and earthling works associated with the cultural operations and harvesting accelerate the erosion reducing the fertile soils into abandoned wasteland (Sharma *et al.*, 2006). However, when the land is being kept fallow for 3-5 years the production level continues. Under the land abundant situation, which is prevailing in the state, the production level for majority of the farmers were not declining below the alarming level because soil fertility usually regenerated by keeping the land fallow for at least 3-5 years. However, improved method of cultivation practices as developed by the ICAR Research Complex for northeast region may enhance the productivity of ginger considerably. Other major production constraints include, lack of knowledge of Integrated Pest Management (IPM), labour constraints particularly during harvesting of crops and seed storage problem of flowers.

Most of the farmers expressed their concern over the wide price fluctuation and price fall is a major marketing problem for all the high value crops including spices. Lack of knowledge on primary grading and standardisation is another problem to realise better prices for the produce. Farmers' expectation from the government is to ensure remunerative prices for their produce to reduce the risk of getting lower price. Farmers expressed the need for training in cultivation of high yielding and hybrid vegetables, training in disease and pest management of vegetables and cultivation of flowers such as anthurium. Earlier studies indicated that lack of approach road was the major problem as reported by 88 per cent of the farmers surveyed in Ri-Bhoi district. Monopoly of commission agents was considered as second major constraint reported by 85 per cent of the farmers. Besides, lack of regulated markets, low price during harvesting were considered to be other major problems in marketing of high value crops in the study area (ICAR-RC-NEHR Annual Report (2005-06)).

Most of the cultivation practices in the study area were observed to be organic only, however, farmers were not receiving price premium for their produce because the process is not formally certified by any accredited certification agency. Therefore, these products are being marketed along with the other conventional products and opportunity of realising price premium is lost. Since there is limited scope to enjoy the economies of scale in this hilly tract by individual farmers, these SHGs can be a good option for expansion of areas and volume of production of these high value crops. This increased production level would be helpful to promote the marketing of these products through encouraging the entry of new private market players whose participation is extremely poor under current situation. The cultivation through these groups may also likely to reduce the organic certification costs for these high value crops (particularly ginger), which is one of the major hassles for promoting organic farming practices in the study area.

Interventions Needed

Meghalaya is suitable for growing a number of high value crops and cultivation of these crops are likely to increase farm income and likely to keep the farming profession remunerative. In hilly tract of Meghalaya, the scope for expansion of area under cultivation is limited. Besides, immobility of labour force, inadequate marketing support and existing land tenure system aggravate the problem of expansion of areas under these crops. Therefore, farmers are observed to be growing a number of high value crops but with very low level of marketable surplus. Under this circumstance, the formation of SHGs to increase the volume of production is a good and timely option to improve the farm production as well as income. Results indicated that the SHGs has motivated the farmers/ growers to grow existing high value crops and also encouraged to introduce new crops. However, increase in terms of net income due to formulation of SHGs was not observed to be significantly different from the growers without being a member of SHGs. Also farmers were uncertain whether or not to continue newly introduced crops (such as broccoli) when the government support (for example, DHO supported through providing seeds and ensuring price to buy back the products) is being withdrawn. Notwithstanding with these weaknesses, in general, it has been observed that the formation of SHGs have been articulating desired enthusiasm and encouragement among the primary producers and gradually they are willing to take up new crop enterprises which may facilitate improvement of farm income. Under this process government may take up initiative to encourage private investment through introducing incentives for their active participation in various marketing functions like grading, processing, storage, packaging, branding and transportation particularly for the export promotion of high value crops such as organic spices, fruits and vegetables and also flowers. Lack of infrastructure and markets is an important impediment in realising the potential of high value agriculture in the region. Local markets for high value food commodities are thin, and transport, cold storage and processing infrastructure is poor (Birthal *et al.*, 2006).

Under the present economic liberalisation process, agricultural marketing system has to be strengthened and restructured to facilitate the improvement of farm income. Farmers should be made aware of and capable of taking advantage arising out from the ongoing marketing innovations and also from various schemes of government and other institutions. Alternative institutional arrangement like contract farming, farmers' companies and new generation cooperatives for coordinating the marketing efforts of small farmers should be evaluated in different social and cultural settings and encouraged for adoption according to social feasibility (Acharya, 2007). This emerging agricultural market situation (like entry of multinational companies in retail business of agriproducts) is likely to influence the share of primary producers either positively or negatively depending on how the farmers can respond to the changing demand situation of various agroproducts. In view of this it is essential to increase the awareness level on the recent changing pattern in Indian agricultural marketing situation among the farmers and making them empowered to take better decision on production and marketing of agricultural commodities. State government should be more proactive to encourage the private players in agricultural commodity marketing. Market innovations like contract farming can be encouraged in the state to facilitate the marketing of these high value crops as well as organic products.

9 Biodiversity and Organic Agriculture

Opportunities and Challenges for Northeast Region

A close relationship exists between organic agriculture and maintenance of biodiversity which is expressed in the rules and regulations that govern certified organic agriculture and by practical experiences gained by the farmers who practise organic agriculture around the world. In developed countries, the role of organic agriculture in the restoration and maintenance of species, varieties and breeds at the risk of extinction have also led many organisers or individuals to adopt this system to save genetic species and ecosystem diversity as well. Conservation efforts, through organic management, are however even more important in centres of diversity of crops. In these areas, cultivation of populations with high genetic variability is an indispensable inheritance of agriculture and as such, food security for future generations (Scialabba *et al.*, 2002). Global population is increasing rapidly and the time span for increasing total population of one billion is getting shorter and shorter. It took 100 years to reach the population of two billion during 1930s, while the addition of five billion of population took hardly 12 years of time (1975-1987). It is estimated that the population explosion will touch to eight billion during 2022 AD. Similar to this upward trend at global level, Indian population growth rate is also high enough and presently supports 16 per cent of world population with the possession of 2.4 per cent of total land area. One of the biggest challenge for the world in general and India in particular is to secure food for this burgeoning population. As per the Food and Agriculture Organization (FAO) estimate, during 2025 world has to feed four persons from a hectare of land which was only two persons/ha during 1950. Considering the inelastic supply nature of arable land, this scenario is clearly ringing the alarm bell for the world food crisis in near future.

To meet the future food demand it needs to produce more food per unit of land for which the varietals improvement of various food crops are extremely important. The success of the new variety development programme is heavily dependent on the availability of suitable plant genetic resources. The genetic resources are continuously used by the scientist for improvement of crops to meet the challenges of increasing food production which has worldwide demands. Thus, the conservation of plant genetic resources has been the focal issue among the agricultural scientists over the past few decades. The erosion of agricultural biodiversity threatens long-term stability and sustainability of agricultural production. The majority of high-yielding varieties (HYVs), particularly hybrids, are not very long living as they tend to loose their viability and productivity, or become increasingly susceptible to pest/disease attack, within few years. This necessitates infusion of fresh genetic material, which is again obtained from traditional varieties or their wild relatives. Though the cost of conservation is quite high, we can not afford to loose any potential and viable landraces, because once a landrace is lost it is lost for ever. As the value of one potential landrace is exceedingly high (not possible to measure exactly as it will depend on its future uses), it must be conserved for future utilisation. The adoption of high yielding, uniform cultivars have led to considerable reduction in the number of genetically viable species of landraces used in agriculture. Many food crop varieties have virtually disappeared from their centres of diversity. There is now an increasing body of evidence that organic agriculture supports and maintain much higher level of biodiversity than conventional farming systems, including species population that have significantly declined. The relationship between organic agriculture and agricultural biodiversity is expressed at the philosophical and theoretical levels in basic principle, standards and regulations that govern organic agriculture. According to the FAO/WHO Codex Alimentarius Commission Guidelines for organic food 'organic agriculture is a holistic production management system which promotes and enhances ecosystem health, including biological cycles and soil biological activity... The primary goal of organic agriculture is to optimize the health and productivity of inter-dependent communities of soil life, plants, animals and people'. Organic agriculture offers a means to address food self-reliance, rural development and nature conservation. The common thread in this ambitious goal is the sustainable use of biodiversity; in terms of both agriculture contribution to biodiversity and biodiversity contribution to agriculture (Scialabba, 2003). Therefore, the organic agriculture system has strong positive correlation with the maintenance of and enhancing biodiversity through *in situ* conservation and sustainable use of genetic resources. Many of the empirical studies around the world has given the satisfactory results and have since become focus of the research institutions.

The northeastern region of India, one of the 18th mega biodiversity centres of the world, has the unique opportunity to supply scarce plant genetic resources and simultaneously has the strong potential to become the leading growing centre for organic food production. The region has a number of advantages, like, minimum use of fertiliser and pesticides and highly favourable climatic condition suitable for growing diverse crop varieties. The theme of the present chapter is to highlight the potentiality of organic agriculture while maintaining the agrobiodiversity in northeastern hill region of India. The discussion has been divided into four sub-sections covering the broader issues; first, why northeastern states should be considered for the conservation of plant genetic resources as well as potential centre for organic food production; second, what is the present status and progress made so far on agriculture front and feasibility of adopting organic agriculture production system; third, what is the present status of biodiversity, particularly agrobiodiversity; and finally, highlight the opportunities, challenges and issues for managing such biodiversity while converting the NER into a centre for organic agriculture production.

Green Revolution Bypassed: A Boon for Northeast Region?

The benefit of Green Revolution in India was spectacular in terms of increase in food grains production by manifolds, which resulted into transforming India into a 'food surplus' from a 'food deficit' country. But this benefit has not been percolated to many parts of the country; rather, it was restricted to few pockets only. The land available for agriculture continues to shrink and face the environmental threats in many ways, e.g., genetic erosion, desertification and other transboundary pollution. Agricultural intensification contributes to over 20 per cent of global atmospheric greenhouse emission, and has affected 70 per cent of all threatened bird species and 49 per cent of the total plant species (Scialabba, 2003). Moreover, these activities though help in reaching higher levels of production but leave many side-effects, such as, indiscriminate use of chemical inputs, aggressive propagation of hybrid seed varieties and uncontrolled mining of water resources, which are the major threats to sustainability of Indian agriculture. The modern practices have resulted into serious problems like, degradation of fertile soils, severe water logging in irrigated areas, and increasing depletion of soil nutrients. The modern agriculture is heavily dependent on utilisation of HYVs, which mostly leads to monoculture and thus replacing the traditional varieties or landraces. So, the introduction of these HYVs itself has become a threat to cause genetic erosion of native/indigeneous crop landraces.

It is now a well accepted fact that the adoption of high-yielding, uniform cultivars of many crops has led to considerable reduction in the numbers of genetically viable species that were used in agriculture. Many food crop varieties have virtually disappeared from their centres of diversity. The NER of India is one of the rare sites where the progress of modern varieties is less and a large number of traditional landraces are being grown and managed sustainably by the farming community. The Green Revolution has virtually bypassed this region and the advancement of modern crop varieties are restricted to a few pockets only. The traditional farming systems are being followed and most of the systems are operating in subsistence to semi-commercial stage. This perceived weakness has tremendous potential to be converted into a unique opportunity arising out from the changed marketing structure, as far as the supply of vulnerable genetic resources as well as production of organic products are concerned. The farming practices in northeast region are not responsive to the market demand, and most of the crop landraces are traded through small-scale informal marketing channels. The primary objective of growing crops (particularly rice) is not for profit maximisation but to attain the household level food security (Mandal *et al.*, 2005). The farmers are continuing to grow a large number of landraces since ages; therefore the opportunity of *in situ*/on farm conservation is quite high and might be feasible with least opportunity cost. The traditional farming systems are highly eco-friendly where local crop varieties are most suitable to grow because these are more resistant to biotic and abiotic stress in comparison with the modern varieties. The non-adoption or slow adoption rate of modern agriculture has helped the farmers of these backward states to maintain and manage the agrobiodiversity sustainably, which has enormous public value for the country. Geographical barrier, non-availability of suitable variety seeds and other inputs, lack of adequate infrastructure, poor extension services and finally highly imperfect market situation has made the region less progressive in agricultural front and in turn resulted into low external input-use farming systems. Perhaps this slow progress in modern agriculture is the biggest asset for the region to be a leading zone organic food producer in the country. The northeast region is said to be, by and large, organic by default or the farmers in this region are *de facto* organic producers. Truly, the farmers of these states are using very nominal (or not at all) amount of inorganic inputs, fertiliser (less than 35 kg/ha) or pesticides. Recently, the use of fertiliser has been started in few pockets of the region, but the level of use is almost negligible, except Manipur (104 kg/ha). The low external input-use status is the biggest strength for the possibility of organic agriculture in this hilly region. The conversion period of inorganic to organic agriculture, is to be much lower and easier in this region as compared to other parts of the country, as the conversion process will be highly supported by the readily available forest biomass and the residues from diverse livestock population. During the transition period the farmers used to face the problem of yield losses which will be also minimum because the farmers are using very low level of inorganic inputs. Organic products usually fetch higher prices, nearly 20-25 per cent, a premium price over conventional products, which may compensate the problem of yield losses during early period of conversion and is expected to increase the level of income.

By choice or by compulsion the farmers of northeast do not make use of synthetic agricultural inputs, rather, they rely on 'natural input' by enhancing biodiversity through *in situ* conservation and sustainable use of genetic resources. Independent of the motives that the farmers have for the adoption of organic agriculture, in every case a marked increase in biodiversity is visible. Several studies showed that organic agriculture supports much higher level of biodiversity than conventional farming system, including species that have significantly declined. Most importantly, diversity is the basic insurance against crop and livestock disease outbreaks (Scialabba *et al.*, 2002). Thus, the maintenance and conservation of agrodiversity can be managed successfully through promotion of organic agriculture, while blending it with the proper scientific knowledge and technology input for the sustainable development of the region.

Loosing Biodiversity: A Threat to Insurance of Global Agriculture

Biologists believe that they are currently witnessing a wave of mass extinction of species but this time the sudden loss of biological diversity is due to human actions (May *et al.*, 1995). The exact rate of extinction is difficult to assess because we do not know how many species there are, however estimates ranges from 3 to 25 per cent of species loss between the mid-1980s to 2015 (Barbier *et al.*, 1994). By taking land for agriculture and other uses mankind is gradually using up the habitats which support biodiversity. The loss of a species is a cost to society for three reasons. First, the species lost may have a direct value in terms of, for instance, genetic material for food crops, or as a source of medicines. Second, a species may play a critical role in maintaining an ecosystem and providing humanity with a range of ecosystem services such as nutrient cycling, and water catchments regulations. Thirdly, a species may have aesthetic and non-use benefits, for instance people like to see 'charismatic' species such as, birds and butterflies in the wild (Virchow, 1999). Since, human activities have increased the rate of extinction; therefore, economics should be involved in findings solutions. Shogren *et al.* (1999) have highlighted the importance of economics to make decision that society has to take for conserving biodiversity. Thus, the economic activity determines the risk and value of a species that faces extinction.

The agricultural biodiversity includes all components of biological diversity of relevance to food and agriculture, including genetic resources for crop and livestock. Agricultural biodiversity provides food, income and material for clothing, shelter and medicine. It also performs ecological services essential to human survival, such as nutrient cycling, pest and disease regulation and pollution control. The risk of species or varieties or landraces or wild relatives of crops, extinction is high in the modern agricultural system as compared to the traditional farming systems practising in northeast region of India. The magnitude of genetic erosion of crop diversity is alarmingly high at national level. For example, 70 per cent to 90 per cent of the total cropped area has been covered under HYVs for rice and wheat respectively. In northern India, Sonalika alone has covered half of the total wheat area under cultivation. In Andhra Pradesh, 95 per cent of the local landraces of rice has been disappeared (Kothari, 1999). Similarly, in Chhattisgarh, prior to 1960 nearly 19,000 rice landraces were cultivated, out of which majority of them are currently lost. The northeastern states are also loosing its precious crop landraces over a period of time but the rate of loss is considerably low as compared to other parts of the country. Studies indicated that there is loss of 10 per cent local rice landraces over a period of 10 years in NE region. However, case study in Tripura showed that the rate of loss in the past 15-20 years period varied from 12.5 per cent to 73.3 per cent (Sharma and Hore, 1990). The rate of loss of rice landraces has been observed to be almost insignificant in Arunachal Pradesh. Though the spread of research varieties implies a possible narrowing of the gene pool but the farmers of the hills are more dependent on their own landraces which was generated from their previous generations. Even they did not discard the method of practice of cultivation, which was adopted by their forefathers. This is possibly due to their taste and preference towards local varieties and to attain food security. Local farmers who inhibit in remote hilly areas, are more conscious about the selection of landraces, which is suitable for specific environment. They have acquired this technique through several decades, based on their experience (Hore, 2005). Rice is the main crop in the region and there are diverse cultivars with useful genes. Of 12,000 landraces of rice, about 6,000 come from the Northeast Himalayas. There is wide variability in the rice germplasm collected from different parts of the northeastern region, but glutinous and japonica forms dominate the endemic types. Some of the cultivars are resistant to diseases and pests and they are tolerant to drought and cold.

The northeast region is rich in primitive and endemic flora which have great economic importance also. Since long past experts have opined that the region can be considered as an active speciation zone, as a result of natural mutation, hybridisation and floral evolution (Kanjilal, 1936; Robinson, 1841; Hooker, 1854 and 1906). The citrus fruits include oranges, limes, lemons and pommelo. About 37 citrus species are cultivated in India. Bhattacharya and Dutta (1951) described 17 species, 52 varieties, and 7 natural hybrids of citrus fruits that come alone from undivided Assam. *Citrus indica*, a primitive and endangered species of citrus and the probable progenitor of cultivated species, grows in the Garo hills. About 100 landraces of citrus have been collected by ICAR. Citrus (*C. medica*) is planted widely in kitchen gardens all throughout the region and has a wide range of variation in fruit (Table 9.1). Orchid is a unique group of plants with a wide range of variability. The main economic importance of orchids lies in their ornamental value, but some species also have a medicinal and/or aromatic value. Of 1,300 species belongs to 163 genera reported in India, more than 600 are from the northeastern Himalayas. Sikkim alone has more than 450 species belonging to 100 genera. Eighteen endemic species are found in Arunachal Pradesh (Hegde, 1984). The orchids are found from tropical to alpine zones. As many as 225 orchid species are rare, 12 endangered, 16 vulnerable, and one species, *Paphiopedilum wardii*, is nearly extinct (Upadhyay and Sundriyal, 1998). Regarding bamboo, out of 136 species available in India, around half of the species are native to the northeast Himalayan region (Kochhar *et al.*, 1992). The commonly available species are *Dendrocalamus hamiltonii, D. hookeri, Bambusa balcoa, B. tulda, B. pallida, B. nutans, Cephalostachyum capitatum*, and *Melocanna baccifera*. The rare and endangered species are *D. hamnei, Teinostachyum helferi*, and *Bambusa mastersii*. There are more than 250 fleshy mushrooms, indigeneous to the northeastern Himalayas, belonging to about 100 species. Many of them are edible. Among medicinal plants, 120 wild species have been reported from this region, out of total 3,000 species found in India. Over 1,120 species of insect pests are found in northeast region which are important from agricultural point of view. Out of which nearly one-fourth of them are having beneficial characteristics and rest are important either as agricultural pest or neutral. Three out of five species of honey bee and four out of five species of silkworm found in India are available in this region. Nearly 15-20 species of butterflies are available having ornamental value and high export potential (Kalaisekhar, 2006). Besides, agrobiodiversity the fish diversity is also mind boggling for this region. Most of them are edible. The ornamental fish available in this region have great commercial value at national or international market. Recent estimates of fish species of northeast region showed 274 potential fish species which belong to 114 genera under 37 families and 10 orders. This includes 265 indigeneous and 9 exotic fish species. Nine exotic fish species have already well adapted with the local environment. The fishery scientists have recorded and established breeding population of some of the species like *Cyprinus carpio*, *Oreochromis mossambicus*, *O. niloticas* and *Clarias gariepinus* from different water bodies of the region. Out of 806 fish species inhabiting freshwater of India, northeast region represents 34 per cent (Mahapatra *et al.*, 2004).

Table 9.1

Distribution of Wild Relatives of Cultivated Crops in India vis-à-vis in the Northeast Indian Himalayan Region

Crop	*Number of Species*		*Remarks*
	Northeast Himalayas	*India*	
Cereals	16	60	
Legumes	6	33	
Fruits	51	109	
Vegetables	27	64	
Oilseeds	1	12	
Fibre crops	5	24	
Spices and condiments	13	27	
Miscellaneous	13	26	
Rice (landraces)	6000	12000	
Citrus	17 (Assam only)	37	100 landraces collected by ICAR
Bamboo	63	136	
Orchids	600	1300	450 in Sikkim alone, 225 rare, 12 endangered and 16 vulnerable
Mushroom	100		
Banana		300	
Fish	274	806 (approx.)	
Medicinal plants	120	3000	
Insects (agriculturally important)	1120 (approx.)	-	25% are beneficial (approx.)
Honey bee	2-3	5	
Silkworm	4	5	
Butterflies	15-20	-	Ornamental value and export potential

Source: Upadhyay and Sundriyal (1998); Dubey *et al.* (2002) and authors' compilation from various publications of ICAR.

Saving Biodiversity through Organic Agriculture: An Option

The northeastern states recently have been identified as the most potential area for organic agriculture in the country and accordingly, policy thrust has been laid for necessary internalisation of the region. Being the 'hot spot' in terms of biodiversity, and now identified as the potential organic agricultural production zone, it is necessary to analyse the possible relationship between these two issues, i.e., biodiversity and organic agriculture. So far, most of our research agenda for this region have been focussed separately either on biodiversity or on organic farming, rather missing the inherent link between these two. Many empirical studies have indicated that the biodiversity and organic agriculture have synergistic effect on each other and therefore, biodiversity can be managed well through organic agriculture. Since, food security and conservation of biodiversity both are extremely important, the ecologically sound sustainable production system is a must, for which probably organic agriculture is the best of option for this region. A success story of Khonoma village can be mentioned to support this view. Khonoma, a remote village in Nagaland is well known in agricultural circles for its sophisticated cultivation techniques. In shifting cultivation, farmers use alder (*Alnus nepalensis*) trees interspersed with the crops. These tress return nitrogen to the soil, thereby helping the land regain fertility rapidly when farmers abandon it to move on to the next plot. The village overlooks a wide valley that has been converted into terraced fields, made with such precision that apparently their productivity has remained stable over centuries. Khonoma is also a home of over 60 varieties of rice, and a considerable diversity of millets, maize, Job's tears, citrus fruits and other crops. All the crops are grown without using chemical fertilisers or applying any chemicals, i.e., produced organically (Kothari, 2005).

The northeast agriculture is characterised by shifting (*Jhum*) cultivation which is accounted for nearly 43 per cent (16.72 lakh ha) of total net sown area (NSA) available for cultivation. This is a traditional farming system being followed over centuries, but causing considerable damage to the natural resources, (such as soil erosion, deforestation etc.) as perceived by the agricultural scientists and policymakers. Nevertheless, the case studies on shifting cultivation revealed a number of good practices in terms of using indigeneous methods, such as soil and water conservation and biodiversity management. This calls for fresh research and getting new insights to the entire gamut of hill farming system. The Director General of International Centre for Integrated Mountain Development (ICIMOD), Dr. James Gabriel Campbell pointed out: 'we only used to think of short term and high productivity of singly monocrops, now we are thinking in terms of organic, sustainable production, and biodiversity' (Chakraborty, 2005). Several studies also highlighted the importance for biodiversity in forest gardens which are maintained in many shifting cultivation systems (De Jong, 1997), as these gardens contain a mixture of productive species and manually occurring trees. Conversion of forest cultivating area to the intensive permanent farming systems with monocropping, will invariably lead to strong reduction in biodiversity.

As long as the region remains sparsely populated with low population density, there will be less human activities on the forest land and minimum threat of loosing biodiversity. However, in view of attaining food security for the region, suitable agrotechniques must be infused with the traditional farming practices. Simultaneously, the agricultural profession should be made more profitable so that labour force engaged in farming improve productivity as well as remain gainfully employed. Perhaps organic agriculture with scientific appropriation is the best answer for this to fulfill both the purposes, i.e., maintaining biodiversity as well as improving the income of the farmers through increasing the agricultural output per unit of area.

Opportunity and Challenges for the Northeast Region

Farmers and forest dwellers are the main users and managers of the natural resources around the world and the world in unison has the responsibility of protecting and conserving plant genetic resources for securing our future food security. There are some 350 biosphere reserves in 85 countries around the globe to protect ecosystems (Scialabba, 2003). The interdependence of countries on plant genetic resource utilisation for food and agriculture is beyond any doubt. Therefore, loosing potential gene pool in country/region must have impact on other countries/regions also. Over two-thirds of developing countries acquire more than half of their crop production from crop domesticated in other regions (Wood, 1988). Only

South Asia, Southeast Asia and West Central Asia depend on food crops originated from other countries for less than 50 per cent of their food production. In all, over 50 per cent of the food production is based on species originated from other regions (Kloppenburg and Kleinman, 1987). Realising the need of the hour selected international codes and agreements expressing the organic agriculture-biodiversity linkages and need for further consideration. The Vignola Declaration and Action Plan on the relationship between nature conservation, biodiversity and organic agriculture of the International Federation of Organic Agriculture Movements (IFOAM), the World Conservation Union (IUCN-WCU) and the World Wide Fund (WWF) for Nature has recognised that, 'organic agriculture is essential for conserving Biodiversity and Nature' and 'asks environmentalists, farmers, politicians, industry and international institutions to support and develop organic agriculture which is ecologically sound agriculture system' (IFOAM, IUCN, WWF, 1999). The Convention on Biological Diversity (CBD) 'encourages the development of technologies and farming practices, that are not only increase productivity, but also arrest various degradation as well as reclaim, rehabilitate, restore and enhance biological diversity. It also monitor adverse effects on sustainable agricultural biodiversity 'such as *inter alia*; organic farming' (CBD, 1996). The CBD Global Strategy for Plant Conservation includes outcome-oriented global targets for 2010 whereby using plant diversity sustainable considers, among others, organic food as an indicator of direct measures to reach the target of '30 per cent plant-based products derived from sources that are sustainable managed' (CBD, 2000).

However, the fundamental question remains that on what account farmers who are managing these diversified unimproved (landraces of traditional varieties) crop varieties, should be encouraged to do so instead of growing the modern varieties and sacrificing their profit earnings. In other words, as the opportunity cost will be higher, the farmers will tend to replace the landraces and will switch over to modern varieties, if proper incentives systems are not instituted. Such, specific situation of managing plant genetic resources for the greater benefit of the society as a whole makes it necessary to discuss the exchange system as well as the property rights of the plant genetic resources. What we call protection of intellectual property rights (IPR). Genetic resources, until recently, have been regarded as the 'common heritage to mankind', which was specified in the CBD. According to this contracting parties reaffirm that the, 'States have sovereign rights over their own biological resources...' (UNEP, 1994). This reaffirmed national sovereignty does not, however, solve the problem or propose any solution for the question of ownership of genetic resources (Virchow, 1999). There has been the development of national and international trade and commercial agreements to determine the properly rights for genetically and biochemically improved resources and, needless to say, India is also member and signatory of many of these organisations such as, International Union for the Protection of New Varieties of Plants (UPOV), Trade-Related Aspects of Intellectual Property Rights (TRIPS) and CBD. Regulations under patent law have successfully been extended to living organisms and subsequently entering into areas of agricultural interest for approximately 20 years. The TRIPS agreement, paving the way for the World Trade Organization (WTO), has been enforced at the international level to harmonise protection laws for all technological inventions including the patenting of plants. Hence, two systems (UPOV and TRIPS) protecting new knowledge presently exists, which advocate the demand side of genetic resources (Virchow, 1999). The International Undertaking on Plant Genetic Resources (IUPGR) has been adopted in strengthening the rights of the suppliers of genetic resources and acting as a counterbalance to the increasing protection of technology resolving out of genetic resources. Although CBD and IUPGR advocate the suppliers of genetic resources in their objective of benefit sharing, TRIPS and UPOV generally support the demand side of the breeders and biotechnology industry. Therefore there is lack in coherence between the different agreements that has emerged, thus complicating the development and adoption of exchange system for plant genetic resources for food and agriculture (PGRFA) (Virchow, 1999).

In view of the above discussions and concern on the maintenance, supply and benefit sharing of the plant genetic resources, it can be stated that northeast region of India has to take several steps ahead to protect their biodiversity for the benefit of the nation in general and the region in particular. To safeguard this plant genetic resources, inventorisation, cataloguing and characterisation of resources both at morphological and molecular level is extremely important. Molecular characterisation will be essential to tackle the problem of biopiracy which might be encountering frequently under the liberalised economy. There is urgent need to create mass awareness regarding the IPR issues among the educationists, administrative people, and researchers and also among the farmers. Appropriate system should be evolved for enforcing the rules and regulation to protect intellectual property rights and ensuring the benefit sharing among the corporations, breeders as well as farmers.

Issues on Biodiversity Management in the Northeast Region

Considering the changing market economy of the world, the biodiversity 'treasure land' of India, the northeast region will have to face multifaceted new challenges, and accordingly the region must get ready for, in before hand, to protect/safeguard the genetic wealth for greater interest of the nation. Identification, cataloguing and characterisation of various crop species, varieties, landraces or wild relatives of crops are extremely important. Keeping updated information on status, utilisation, magnitude of threat, rate of loss of the plant genetic resources for food and agriculture is essential for evolving suitable institutional arrangements and subsequently required to be substantiated by technological imperatives. Formulation of appropriate strategy for conservation of genetic resources is to be framed through extensive and in-depth site/location specific micro-level case studies and subsequently converting these studies into macro framework to derive suitable policies. The choice of conservation strategy mainly depends on the material (species/habitat) to be conserved, i.e., the life cycle, mode of reproduction and the ecological status. Therefore, two major approaches for crop diversity conservation i.e., *in situ* and *ex situ* conservation. *Ex situ* conservation refers to the conservation of germplasm away from its natural habitat. This strategy is particularly important for crop gene pools, and can be achieved by propagating/maintaining plants in genetic resource centre, botanical gardens, tissue culture, and repository or in seed gene banks (OECD, 1999). However, the *ex situ* conservation strategy alone can not solve the problem of loosing biodiversity; rather, it should be considered as the complementary with the *in situ* conservation. Because *ex situ* conservation has certain limitations, such as: (i) it does not maintain evolutionary process that created the crop germplasm diversity;

(ii) many important species are underrepresented because of the recalcitrant nature of the seeds; and (iii) genetic shifts or alterations cannot be ruled out due to inappropriate storage conditions. The crop genetic resources are not exposed to natural or artificial pressure and therefore no further chance exist for evolution or adaptation. Considering these weaknesses of *ex situ* conservation, side-by-side the alternative approach, *in situ* conservation strategy may be adopted to protect the genetic erosion of these valuable germplasm. *In situ* conservation is the conservation of ecosystems and natural habitats and the maintenance and recovery of viable populations of species in their surroundings while in case of domesticated or cultivated species, in the surroundings, where they have developed their distinctive properties (UNEP, 1992 and 1995). The CBD has given highest priority to this approach of conservation, which includes species protected in the wild as well as the landraces (i.e., cultivars adapted to the local climate, soil, pests as well as satisfy the taste of local people and other cultivated forms maintained by farmers (Primack, 1993). This also includes the preservation of indigeneous knowledge (social, cultural and religion status), agro-ecosystems and other wild cultivars (UNEP, 1992). Thus, *in situ* conservation enables to preserve evolutionary processes that generate new germplasm diversity under natural conditions. It maintains the important field laboratories for crop biology and biogeography which serves as a continuous source of germplasm also for *ex situ* conservation. Further, for those countries, which have abundant crop germplasm resources, it provides an important option for conservation with wider participation.

Therefore, *ex situ* and *in situ* conservation strategies should be viewed as complementary to each other. The *in situ* 'on farm' conservation strategy may be quite suitable and feasible to conserve plant genetic resources/ diversity in northeast region, because the agriculture is still continuing at subsistence to semi-commercial stage, where a large number of traditional crop varieties are being cultivated. The most important question is why the farmers should be encouraged to grow landraces in addition to grow modern crop varieties, henceforth, loosing the opportunity of profit earnings. So, the concerned issue is to analyse the opportunity cost of growing traditional crop varieties over modern varieties. It is easily understandable that as the opportunity cost of growing traditional varieties will be higher, the farmers will tend to replace the landraces and switch over to modern varieties at a faster rate. Since, in northeast region this opportunity cost is considerably low as compared to other parts of India, the rate of loss of crop diversity will also be low (at least in short run), and therefore, the *in situ* conservation strategy may be a viable option. However, this *in situ* conservation may not be a sustainable proposition in the long run for this region unless some appropriate incentive mechanism is instituted through suitable policy dimension. As the biodiversity is a global public good therefore, cost of conservation should be borne by all countries and not only those who host the biodiversity. Thus, proper international institutional arrangements are essential for conservation of biological diversity, the sustainable use of its components, and the fair and equitable sharing of benefits from the use of genetic resources. This is the unique opportunity for this underprivileged northeast region of India, which is arising under the emerging liberalised economy. International policy is already into action, such as the Global Environmental Facility (GEF), which provides funding for the projects on biodiversity protection from developed countries to developing countries. The principle behind this funding is that it pays for incremental costs of projects which provide additional global benefits.

There is a growing need to translate international policy into national and local policies which can provide additional protection. However, while formulating the suitable strategy for biodiversity conservation, policy-makers have to take care of the wellbeing of the farmers' community so that the pace of agricultural development should not be slowed down in northeast region. In fact, the policy thrust should be put forth to minimise the magnitude of opportunity forgone due to conservation of traditional varieties/landraces which are supposed to be utilised in future varietals development programme. The policies have to look into ecologically sound sustainable agriculture comprising two-way opportunities, i.e., feasibility of biodiversity conservation as well as improvement of farmers' income. Perhaps, organic agricultural is the best option to attain these twin objectives particularly for this hilly region. The organic agriculture has its ability to produce high value commodities, for which the demand is increasing both nationally and internationally, as well as it will help to maintain biodiversity at all levels. However, it is logical to assume that in wild areas, organic agriculture may disturb natural habitats by the very fact of human intervention. In any case, it offers an important step towards a solution to many of the threats that conventional agriculture has on biodiversity and this should be considered simply as the most appropriate starting point from which additional conservation needs, where they exist or can be built (Scialabba, 2003). Considering the organic agriculture as the best option for managing biodiversity as well as improving farm income in northeast region, however, many issues are to be resolved while implementing such policy initiatives. Some of the important issues related to conservation of plant genetic resources will be: (i) species/landraces/varieties to be conserved and their cost; (ii) identification of different significant factors that determine that the farmers in this region will continue to grow landraces; (iii) deciding the stage of intervention for adoption of *in situ* conservation strategy; and (iv) kind of effective support mechanism to be provided in order to maintain the *in situ*/on farm conservation of various potentially viable landraces of crops? Apart from these issues on conservation, some important issues, relevant for northeast region are to be resolved on organic agriculture. These are: (i) identification of the crops can be grown through organic agriculture together, with economic feasibility, and (ii) exploration of different marketing channels of these products so that farmers' income can be stabilised. Therefore, to achieve the goal of biodiversity management through organic agriculture, several multidisciplinary researches are to be carried out to formulate the proper policy framework. In-depth analysis are required for ethnobotanical and socioeconomic studies to understand the farmers' knowledge. Selection criteria of germplasm and their innovative management are important to know. Such indigeneous knowledge also warrants due share and benefit when it is applicable to advance technology. Dynamics of the local landraces and farmers' varieties needs to be understood in detail and, simultaneously, appropriate scientific know-how should be infused with the traditional agricultural practices for optimum exploitation of natural resources and ensuring the synergistic effect of biodiversity-organic agriculture in northeast region.

10 Emerging Issues on Agricultural Development in the Northeast Region

Fallacy of Input-Output Ratio

Mostly the profitability of different farming system models at ICAR Complex, Barapani have been judged and compared with shifting cultivation considering purely the input-output ratios only (ICAR, 2005-06) which is not a correct evaluation criterion as far as viability of the system is concerned. Because, input-output ratio does not account the time value of money and also involves only the annual operation and maintenance cost, completely avoiding the investment made on the land development and infrastructure creation. Inclusions of this capital investment in financial viability analysis reveal that the proposed farming system model may be a viable proposition but not be highly attractive one. The farmers in the hilly region are mostly having poor resource base and low economic status with limited investment capacity. Moreover, the abject poverty creates a preference pattern where future income is discounted at a very high rate. In fact, it can be argued, poverty is such a serious problem that all consideration of the future is totally absent in their calculations (Datta and Singh, 2004). The irony is that, the obsession with maximising production continues and yield levels are still the primary discriminating criterion. Profit maximisation, efficiency of production, risk minimisation are important objectives for farmers but these are not used as evaluating criteria (Jha, 2001). In this situation farmers are not very much keen to (or capable of) investing capital on land development in expectation of future returns. Thus, the profitability analysis of farming systems should be made more critically taking into account time value of money of capital investment required as well as operating and maintenance cost.

Yield Variability and Risk Bearing Ability

The different crop activities at farming system research at ICAR Complex, Barapani shows wide yield variability, e.g., maize (44%), ginger (79%), radish (83%), groundnut (33%) and french bean (83%), which implied high risk association along with these crop activities (Ibid.). This variability has been observed in research farm (agri-horti-silvi-pastoral farming system) in spite of giving all necessary and timely inputs along with scientific support. For example, a case study in Sasatgre village, West Khasi hills, showed that *jhumias* learned better orchard (orange) management to promote better cash returns and they suspended their annual *jhumming*, the tiresome shifting cultivation, and instead, decided to invest their labour—the scarce and only available capital—into developing orchards, after a bumper harvest of orange in the previous year fetched them an unexpected bounty of hard cash. They get some extra income from orange orchards, when there is fruiting, which all these years had been erratic. It is like gambling with the natural elements and the decision was a disaster as most of the young fruits dropped pre-maturely. Now, again villagers have returned to their traditional *jhumming* with the view that, you may not get sufficient food from the *jhum* field but it never returns you empty-handed (Chakraborty, 2005). Keeping such situation in reality, the major concern is how far the resource-poor farmers can afford this risk where the farmers are mostly risk averters.

Profitability *versus* Subsistence

The objective of farming system research is to transform shifting cultivation into a permanent agricultural system through ensuring higher secured earning. But the real concern of the farming community is food security rather than profitability due to poor market linkages and hindrance to free mobilisation of surplus output. Thus, the underlying objective of farming system research (profitability) and objective of shifting cultivation (subsistence) is completely diverging, in which a balance is possible to achieve only when different pulling and pushing factors (such as, road linkages, value addition in agricultural product like organic farming) are properly identified and fulfilled. Still then the farming system research needs to be reoriented aiming at improvement of present shifting cultivation, rather to replace it. The reorientation may possibly drive the present subsistence level of farming to semi-progressive level and in turn to commercial level through improving the investment capacity of the farmers. One of the major arguments for prohibiting shifting cultivation is the low productivity of crops. It is true that the productivity of crops grown in shifting cultivation areas is quite low when we consider the crop productivity individually. But the overall productivity (mixed cropping) from shifting cultivation method might be much higher. However, precise quantification of total productivity in shifting cultivation is very difficult and it requires separate methodology to be evolved. Moreover, shifting cultivators often have poor access to agrochemicals because of poverty, inadequate extension systems and a general disregard of their farming practices by authorities. Appropriate management solutions should be made available to farmers, including improved upland crop varieties, and more effective use of fallow vegetation by incorporating useful species (Cairns and Garrity, 1999).

Biodiversity and Shifting Cultivation

Conservation of biodiversity is one of the globally important and critical issues for world agricultural development, because the erosion of agricultural biodiversity threatens long-term stability and sustainability of agricultural production. The northeastern hilly region of India is one of the world's 18 hot spot areas as far as biodiversity is concerned. The traditional farming systems are being followed and most of the farming practices are operating in subsistence stage, maintaining huge biodiversities, which have enormous public value for the society. Some case studies on shifting cultivation indicated few good practices in terms of using indigeneous knowledge including soil and water conservation and biodiversity management. There is a need to look into the entire hill farming system in these perspectives. Dr. J. Gabriel Campbell of International Center for Integrated Mountain Development (ICIMOD) has opined that there is need to think of hill farming system in terms of organic, sustainable production, and biodiversity. Continuous conversion of forest or shifting cultivating areas into the intensive permanent farming systems with few crops may lead to loss of valuable biodiversity. Thus, while suggesting the permanent agricultural system in *jhum* areas we must consider this biodiversity issue keeping in mind the future agricultural development.

Missing Link in Identification of Farm Level Needs

Hilly tribal societies have fashioned intermediate, often criss-crossing institutions, such as the household, extended-family and kinship networks; civic, commercial, and religious associations; charities; production units; and various layers of what is known as government. Each serves functions at which the others are not so good. They differ not only in terms of the emotional bonds that connect members, but also in regard to the information channels that serves them, the kinds of agreements that bind them, and the investment outlays and severance costs that help sustain them. Their elucidation, in particular our increased understanding of their strengths and weaknesses, has been the most important to compel the process of technological intervention for any economic growth and development. The challenge is to identify specific agricultural and rural needs, and to focus investment in areas where the greatest impact on food security and poverty will be achieved. This is made possible through developing an understanding of the local factors and linkages found in the wide range of rural locations subject to differing socioeconomic and ecological conditions. One of the important issue is that the scientists are well aware and familiar of the physical and biological parameters of the system—soils, climate, growing conditions, etc., but what they generally lack is an understanding of the socioeconomic factors which also condition behaviour, both at micro and macro-levels (Jha, 2001). Only agricultural development will not help to increase farm income but more importantly by creating demand for non-tradable goods, namely services and local products. It is this indirect effect on demand—and the associated employment creation—in the non-farm sector of rural areas and market towns, that appears to be the main contributing factor to the reduction of rural poverty. Dr. M.S. Swaminathan said in an interview with reporter of Grassroots Options, '*jhumming* is neither good nor bad, depending upon how you do the *jhumming*. If we still have an opportunity for regeneration of forest and soil fertility then it's alright; otherwise, we can introduce new interventions for *jhumming*, which can eliminate its negative aspects, what I call, ecologically and socially sustainable *jhum* farming and it can be done.'

Several lessons and these issues are important for which we must find a paradigm shift to find sensible solution. First of all we need to have more knowledge, understanding and better insights regarding why farmers do what they do in the shifting cultivation method. More analysis and conscious efforts are required to look for interactions of shifting cultivators with various natural resource components. Special focus is needed to understand their existing socioeconomic conditions and status of resource endowment existing in an almost close economy, where market-driven forces are almost absent. Reorientation of the objectives of farming system research is essential aiming at improvement of prevailing *jhum* cultivation practices, rather not to weaning it away completely. Different pulling and pushing factors of agricultural economic development must be identified properly and accordingly differentiated strategy should be suggested in which farmers' needs/aspirations and understanding of their risk preference, family requirement and resource flows should be given priority. It is also necessary to understand the economic rationality behind the old age farming system practices of the tribal farmers. Their needs directly reflected their landholdings. Due to undulated topography, plot size is very small and scattered; they can not think to enjoy the economies of scale as cost of production is comparatively high and due to insufficient storing capacity of rain water in spite of receiving very high rainfall. They have acquired vast experience from natural wisdom on the crop choice, like growing crops which are sensitive to water stagnation (e.g., pineapple, ginger) and suitable to grow it along the slope, instead of across the slopes. Their knowledge and experience must be incorporated for redefining different levels of farming system approach.

Agricultural sector in northeastern states is experiencing the transition phase, slowly shifting out from traditionally low-income agriculture, but yet to reach the stage of pre-conditions for take-off into self-sustaining growth. Under this stage of economic transition, various sub-sectors particularly under agriculture and allied activities showed high growth prospects, when stimulated through policy initiatives. But the real challenge remains to sustain this growth rates in long run. Under this traditional economy, it is difficult but extremely important to prioritise the sub-sectors which have short-run as well as long-run growth prospects. Sometimes, some of the sub-sectors may even show misleading growth rate pretending to have high or low potential but actually are not, particularly due to the non-availability of reliable database (e.g., growth rate for food grains production in Mizoram can be observed to be as high as greater than

10 per cent when calculated for the period of 1984-85 to 1997-98), emphasising the need for credible database for various sectors. Capacity building and empowering with technical knowledge might be given top priority for the human resource development in this region. The traditional family farmers resist change towards modern technologies and it is necessary to analyse whether their behaviour is stubborn and irrational (e.g., low adoption rate for high yielding variety seeds for rice cultivation) or they are acting rationally within the context of their particular economic environment, and the agrotechnologies should be appropriated accordingly to increase the adoption level of these innovations. Efficient utilisation of plan outlays after prioritising the sub-sectors and minimising the leakages should be emphasised for self-propagation of growth. The focus should be to increase the efficiency of production along with the volume of production; therefore, returns to investment should be the criteria for flow of fund to various projects. Enhancing the productivity of women labour engaged in agricultural activities needs to be looked into more positively (e.g., organising training for women groups at their own villages, giving preferential treatment for women group in obtaining institutional credit etc.). Strict monitoring and evaluation should be followed for implementation of agricultural developmental project. The performance of output growth for various sub-sectors (crop, fruits, vegetables, spice, fishery, livestock, etc.) under agricultural sector indicated that combination of these activities will facilitate to augment the farm income rather relying on any single activities, which is true even at farm household level, thereby suggesting promotion of integrated farming system approach at massive scale in this region. Expansion of market through horizontal integration among the primary producers to increase the volume of marketable surplus and then, vertical integration for value addition of these products is extremely important to realise the potential of this region. The promising areas like expansion of meat production, increasing fish output, growing off-season vegetables, fruits, spices and condiments and promoting organic products should be given priority. Market should not be viewed as 'master' rather; it should be used as our 'servant' and demand creation (e.g., organic rice) for the products from northeastern states should be done with innovations and appropriate strategic planning (e.g. expansion of trade with neighbouring countries through establishing formal trade point at border areas). The volume of marketed surplus from agricultural production must

be increased and value addition should be done at least at primary level to make the product price competitive at domestic as well as international market. The process of economic growth and development for the northeastern states is primarily dependent on agriculture sector and thereby, development planning should continue to follow agriculture-led growth.

11 Policy Implications

Farming systems in northeastern states are complex and characterised with high degree of risk and uncertainty. At household level, particularly in upland conditions, the farming systems are highly diverse with multiple crops (mixed cropping) but at aggregate level the cropping pattern region is skewed to paddy only. The region is highly suitable for growing a number of high value horticultural crops which have great demand in domestic as well as in international market. Improvement of farming system through selection and promotion of suitable crop mix in compliance with the existing socioeconomic as well as topographical situations would certainly enhance farm income. There is a growing need for prioritisation of sub-sectors with potential of short-term as well as long-term growth prospects and framing strategies to sustain these growths through appropriate institutional support and public-private partnership (PPP).

Agricultural markets and value chain analysis may be given special priority to provide the desired impetus for growing high value crops particularly fruits, spices, condiments and flowers. The most important step is the horizontal expansion to increase marketable surplus and vertical integration for value addition of the farm harvest. Extensive database on volume of production and cost of cultivation and ensuring remunerative price based on estimated cost of production. Public sector responsibility in terms of establishing separate price commission can be thought of for the region.

Diversification plan for the whole region based on the macro-level data may not be effective because of prevailing high degree of heterogeneity in cultivatable areas and restriction of growing alternative crops across the topography of the region. Therefore, planning crop diversification through extensive micro-level studies particularly at the upland situations of the farming land is essential.

Organic farming and biodiversity has a close relationship particularly with the hill farming system of the region. Management of agrobiodiversity through organic agriculture may be a viable option for the region to achieve the twin objectives like biodiversity conservation as well as promotion of organic agriculture. Opportunities can be harnessed through popularising organic farming and utilisation of biodiversity resources to facilitate rural entrepreneurship building as well as improving farm income. Product positioning of organic produce, in different niche market (both domestic and international) may be successful through promotion of contract farming particularly for high value horticultural crops, organic foods and meat.

Establishment of market intelligence system for timely dissemination of information regarding day-to-day prices at various market points is relevant to the northeast region. Private investment may be encouraged and various incentives may be provided to establish model agricultural, horticultural, ornamental fisheries and organic food production including meat at district level through cluster approach. Once the backward and forward linkages are established successfully even for a small scale of produce, it will subsequently be helpful to increase the volume of production and agricultural trade in long run.

Finally, the return to investment will certainly depend on how efficiently the amount is being spent. Free and fair environment must be provided at the regional level to attract adequate private investment. Ensuring security needs for the investment in agriculture sector for which state government must play a proactive role by resolving various conflicting and complicated social issues through discussion with the local people.

References

Acharya, S.S. (2007). "Agricultural Marketing and External Trade: Infrastructure and Policy Needs for XI Five Year Plan", *Indian Journal of Agricultural Marketing* 21(2): 141-156.

Anon (1992) *Agro-climatic Planning for Agricultural Development in Meghalaya*. Working Group, Zonal Planning Team, Eastern Himalayan Region. Jorhat: AAU.

Baruah, B.K. and A.K. Sarma (1990). *Agricultural Financing in North Eastern India with Special Reference to Commercial Banks: Some Issues*. Jorhat, Assam: Department of Agricultural Economics, Assam Agricultural University.

Barbier, E.B., J.C. Burgess and C. Folke (1994). *Paradise Lost? The Ecological Economics of Biodiversity*. London: Earthscan.

Barah, B.C. and Susil Pandey (2005). "National Symposium—A Report on Rainfed Rice Production Systems in Eastern India: An On Farm Diagnosis and Policy Alternatives", *Indian Journal of Agricultural Economics* 60(1): 110-136 (January-March).

Bhattacharya, S.C. and S. Dutta (1951). "Citrus Varieties of Assam", *Indian Journal of Genetics and Plant Breeding* 11(1): 57-62.

Bhowmick, B.C., B.C. Barah, Shushil Pandey and N. Barthakur (2005). "Changing Pattern of Rice Production Systems and Technology in Assam—A Spatio-Temporal Analysis of Performance and Prospects", *Policy Paper* 22. New Delhi: National Centre for Agricultural Economics and Policy Research (NCAP).

Birthal, P.S., A.K. Jha, P.K. Joshi and D.K. Singh (2006). "Agricultural Diversification in North Eastern Region of India: Implications for Growth and Equity", *Indian Journal of Agricultural Economics* 61(3): 328-340.

Cairns, M. and D.P. Gariity (1999). "Improving Shifting Cultivation in Southeast Asia by Building on Indigenous Fallow Management Strategies", *Agroforestry System* 47: 37-48.

CBD (Convention on Biological Diversity) (1996). Conference of Parties Decision III/11 on *Conservation and Sustainable Use of Agricultural Biological Diversity. http://www.fao.org/DOCREP/005/AD090E/ad090e01.htm*

———. (2000). Conference of Parties Decision V/9, Section C, Target 12. *http://www.fao.org/documents*

Chakraborty, S.K. (2005). "Protect Jhumlands, Jhumia Rights", *Grassroots Options*: 24-26 (Spring).

Chawii, Lian. (2001), "Old as the Hills", *Down to Earth*: 23-33, May 31.

Census of India (2001). *Census Data 2001: General Population Tables*. Ministry of Home Affairs, Government of India. *http://www.censusindia.gov.in/*

Central Statistical Organisation (2003). *Statistical Abstract-2003*. New Delhi: Ministry of Statistics and Programme Implementation. *www.mospi.nic.in*

———. *Net State Domestic Product (1993-94 Series)*. New Delhi: National Accounts Division. Ministry of Statistics and Programme Implementation. *www.mospi.gov.in*

Centre for Monitoring Indian Economy Private Limited (CMIE) (1997). *Agriculture*. Mumbai.

———. (1998). *Agriculture*. Mumbai.

Datta, Samar K., Gurded Singh and Milindo Chakrabarti (2002). "Ginger and Ginger Products: Management of Marketing and Exports with Special Reference to the Eastern Himalayan Region", *CMA Monograph Series* No. 193. Ahmedabad: Centre for Management in Agriculture, Indian Institute of Management.

Datta, K.K., A.K. Tripathi, Med Ram Verma and A.K. Khan (2003). *Annual Report of ICAR Complex for NEH*. Umroi Road, Umiam, Meghalaya.

Datta, K.K. and S.B. Singh (2004). "Analysis of Traditional Agricultural Practices of NEH region", in B.P. Bhatt, K.M. Bujarbaruah, Y.P. Sharma and Patiram (eds.), *Proc. Approaches for Increasing Agricultural Productivity in Hill and Mountain Ecosystem.* Umiam, Meghalaya: ICAR Research Complex for NEH Region. pp.383-290.

Datta, K.K., S. Mandal and Med Ram Verma (2005). *Annual Report of ICAR Complex for NEH*. Umroi Road, Umiam, Meghalaya.

Datta, K.K., S. Mandal, A.K. Tripathi, S.B. Singh, M.R. Verma and S. Mohanty (2007). "Retrospect and Prospect of Food Security in North Eastern Hilly Region of India", *Agricultural Situation in India* LXIV(5): 37-49

De Jong, W. (1997). "Developing Swidden Agriculture and the Threat of Biodiversity Loss", *Agriculture, Ecosystems & Environment* 62 (2-3): 181-97. April.

Dubey A.K., Dinesh Babu and Y.S. Yadav (2002). "Status and Prospects of Horticulture in NEH Region", *ENVIS Bulletin: Himalayan Ecology and Development* 10(2).

Export-Import Bank of India (2006). *Agri Export Advantage* V(II). *www.eximbankagro.com*

FAOSTAT (2006). *www.fao.org*

FAO (2005). *Livestock Sector in Brief*. India. *www.fao.org*

Government of India (1983). *Report of the Task Force on Shifting Cultivation.* New Delhi: Ministry of Agriculture.

———. (2000). *Basic Statistics of North Eastern Region 2000*. Shillong: North Eastern Council, Ministry of Home Affairs.

———. (2003). *17th Indian Livestock Census 2003: District-wise*. New Delhi: Directorate of Animal Husbandry and Dairying and Fisheries, Ministry of Agriculture. *http://dms.nic.in/ami/home.htm*

———. (2004). *Agricultural Statistics at a Glance: 2004*. New Delhi: Directorate of Economics and Statistics, Ministry of Agriculture.

———. (2005a). *Annual Report 2004-05*. New Delhi: Department of Animal Husbandry and Dairying, Ministry of Agriculture.

———. (2005b). *Economic Survey 2004-05*. New Delhi: Ministry of Finance. pp.178-180.

———. (2008). *Achievements of Horticulture Technology Mission during the Period 2001-02 to 2005-06.* New Delhi: Department of Agriculture and Cooperation, Ministry of Agriculture. *www. http://dacnet.nic.in/techmissionscheme/*

———. (2009). *India State of Forest Report 2009.* Forest Survey of India, Ministry of Environment and Forest.

———. (various dates). *Economic Survey* (various issues). New Delhi: Ministry of Finance.

Hegde, S.N. (1984). *Orchids of Arunachal Pradesh*. Itanagar: Department of Forest, Government of Arunachal Pradesh.

Hooker, J.D. (1854). *Himalayan Journals or Notes of a Naturalist in Bengal, the Sikkim and Nepal Himalayas, the Khasia Mountains*. London: Murray.

———. (1906). *A Sketch of Flora of British India*. London: Reeve & Co.

Hore, D.K. (2005). "Rice Diversity Collection, Conservation and Management in Northeastern India", *Genetic Resources and Crop Evolution* 52(8): 1129-1140.

ICAR-RC-NEHR (2005-06). *Annual Report 2005-06*. Published by the Director, ICAR Research Complex for NEH Region, Umroi Road, Umiam, Meghalaya.

IFOAM, IUCN, WWF (1999). International Meeting on the *Relationship Between Nature Conservation, Biodiversity and Organic Agriculture. http://www.fao.org/documents*

Jha, Dayanath. (2001). "Farming System Research: False Starts and Some Thoughts", *Foundation Day Lecture*. March 1. Karnal: Central Soil Salinity Research Institute.

Jha, D. and S.K. Pandey (2005). *Resource Allocation for Agricultural Research: An Emperical Description*. National Centre for Agricultural Economics and Policy Research/NCAP.

Kanjilal, U.N., A. Das, P.C. Kanjilal and R.N. De (1936) *Flora of Assam*. Vol. I-VII. Calcutta: Prabashi Press.

Kalaisekhar, A. (2006). Personal Communication. Umiam, Meghalaya: Division of Entomology, ICAR Research Complex for NEH Region.

Kloppenburg, Jack R. and Daniel Kleinman Lee (1987). "Plant Germplasm Controversy: Analysing Empirically the Distribution of the World's Plant Genetic Resources", *Bioscience* 37:190-198.

Kochhar, R.K. Arora and E.R. Nayar (1992). "Bamboo in NE Region", in *ICAR Bulletin*. Barapani, Meghalaya: ICAR Research Complex for NEH region.

Kothari, Ashish (1999). "Agro-Biodiversity: The Future of India's Agriculture", Article for *Maharashtra Council of Agricultural Education and Research (MCAER) Book*. February 7.

———. (2005). "The Khonoma Magic", *The Hindu Survey of Environment, 2005:* 125-129. Chennai.

Lerman, R.I. and S. Yitzhaki. (1985). "Income Inequality Effects by Income Source: A New Approach and Applications to the United States", *The Review of Economics and Statistics* 67: 151-56.

Mahapatra, B.K., K. Vinod and B.K. Mandal (2004). "Fish Biodiversity of North Eastern India with a note on their Sustainable Utilisation", *Environment and Ecology* 22 (Spl–1): 56-63.

Mahapatra, B.K., K. Vinod, B.K. Mandal and K.M. Bujarbaruah (2006). "Ornamental Fisheries in North Eastern India", *Research Bulletin* No. 49. Umiam, Meghalaya: ICAR Research Complex for NEH Region.

Mandal, Subhasis, K.K. Datta, Suchitra Mohanty and B.K. Mahapatra (2005). "Risk of Financing Agriculture in the North East Hill Region of India with Special Reference to Meghalaya", *Agricultural Economic Research Review* 18: 87-102.

May, R.M., J.H. Lawton and N.E. Stork (1995). "Assessing Extinction Rates", in J.H. Lawton and R.M. May (eds.), *Extinction Rates*. Oxford: Oxford University Press.

Mishra, A.K. and S.K. Gupta (1998). National Seminar on *Land and Water Resources Management for Sustainable Hill Ecosystem*. ICAR Research Complex, Umiam. August 8-10.

Mohan, Rakesh (2003). "Economic Development of the North East Region: Some Reflections", *Reserve Bank of India Bulletin*. December.

———. (2004). "Agricultural Credit in India: Status Issues and Future Agenda", *Reserve Bank of India Bulletin*. November.

Mohanty, Suchitra and T. Haque (2003). "Regional Disparities in the Flow of Institutional Credit in India", *Journal of Rural Development* 22(1): 79-90.

National Academy of Agricultural Sciences (2001). "Hi-Tech Horticulture in India", *Policy Paper 13*. New Delhi, India.

National Bank for Agriculture and Rural Development (2005). *State Focus Paper* (Meghalaya). Regional Office, Shillong.

National Horticulture Board (2006). *Indian Horticulture Database 2006*. New Delhi: Ministry of Agriculture, Government of India. *http://nhb.gov.in/*

National Sample Survey Organisation/NSSO (2006). *Some Aspects of Operational Land Holdings in India*, NSS 59th Round. Ministry of Statistics and Programme Implementation, Government of India.

———. (2006). *Level and Pattern of Consumer Expenditure 2004-05*. NSS 61st Round. Ministry of Statistics and Programme Implementation, Government of India.

OECD (1999). *Handbook of Incentives Measures for Biodiversity: Design and Implementation*.

Pandey, U.K. (1990). *An Introduction to Agricultural Finance*. New Delhi: Kalyani Publishers.

Planning Commission, *Ninth Five-Year Plan (1997-2002)-Outlays and Expenditures*, Government of India. *www.planningcommission.nic.in*

Primack, R.B. (1993). *Essentials of Conservation Biology*. USA: Sinaurer Associate Inc.

Ramakrishnan, P.S., O.P. Toky, B.K. Mishra and K.G. Saxena (1981) "Slash and Burn Agriculture in North-Eastern India", in H.A. Mooney, T.M. Bonnicksen, N.L. Christensen, J.E. Lotan and W.A. Reiners (eds.), *Fire Regimes and Ecosystem Properties*. pp.570-86. Honolulu, Hawaii: USDA Forest Serv. Gen. Tech. Rep. WO-26.

Ramaswamy, C., T.R. Shanmugam and D. Suresh Kumar (1996). "Constraints to Higher Yields in Different Rice Production Environments and Prioritisation in Southern India", in R.E. Evenson, R.W. Herdt and M. Hossain (eds.), *Rice Research in Asia: Progress and Priorities*. Wallingford, Oxford, UK: CAB International. pp.145-160.

Reserve Bank of India (1998). *Report of the High Level Committee on Agricultural Credit through Commercial Bank* (Chairman: Shri R. V. Gupta).

———. (2005). *Handbook of Statistics on Indian Economy. www.rbi.org.in*

———. (2005). *Handbook of Statistics on Indian Economy. www.rbi.org.in*

———. (2005). *National Income, Saving and Employment. www.rbi.org.in*

Robinson, W. (1841). *A Descriptive Account of Assam*. Delhi: Sanskaran Prakashan.

Roy, B.C. and K.K. Datta (1999). "Rice-Wheat System in Haryana: Prioritising Production Constraints and Implication for Future Research", *Indian Journal of Agricultural Economics* 55(4): 671-682.

Sharma, Y.P., A. Kumar, D.S. Yadav, B. Sasikumar, A.K. Tripathi, M. Anandraj, S.K. Sanwal and V.A. Parthasarathy (2006). "Production and Marketing of Ginger", *Technical Bulletin* No. 24. Barapani, Meghalaya: ICAR Research Complex for NEH Region. March.

Scialabba, Nadia El-Hege (2003). "Organic Agriculture: The Challenge of Sustaining Food Production While Enhancing Biodiversity", United Nations Thematic Group, Sub-Group Meeting on *Wildlife, Biodiversity and Organic Agriculture*, Ankara, Turkey, April 15-16. Rome: Food and Agriculture Organization of the United Nations. *zzyx.ucsc.edu/casfs/instruction/ esa/download/10-Biodiversity*

Scialabba, Nadia El-Hege, Cristina Grandi and Christina Henatsch (2002). *Organic Agriculture and Genetic Resources for Food and Agriculture*. Rome, Italy: FAO. *www.fao.org/docrep/005/AC784E/AC784E00.htm*

Sharma, B.D. and D.K. Hore (1990) "Rice Germplasm Collection in Tripura State", *Indian Journal of Plant Genetic Resources* 3(2): 71-74.

Shogren, J.F., T. Tschirart, A.W. Anderson, S.R. Ando, D. Beissinger, G.M. Brookshire, D. Brown Jr., R. Coursey, S. Innes, M. Meyer and S. Polasky (1999). "Why Economics Matters for Endangered Species Protection and the ESA", *Conservation Biology* 13(6): 1257-1261.

Singh, Manmohan (2005). "North East 2020", *Yojana*: 2-6 (December).

Singh, S.B., K.K. Datta and S.V. Ngachan (2003). "Economic Analysis of Shifting Cultivation", in B.P. Bhatt, K.M. Bujarbaruah, Y.P. Sharma and Patiram (eds.), *Proc. Approaches for Increasing Agricultural Productivity in Hill and Mountain Ecosystem*. Umiam, Meghalaya: ICAR Research Complex for NEH Region. pp.323-330.

Spices Board India (2007). *Spice/State-wise Area and Production*. New Delhi: Ministry of Commerce and Industry, Government of India. *http://www.indianspices.com/*

Thirwall, A.P. (1999). *Growth and Development* (6th Edition). London: Macmillan Press Ltd.

Todaro, Michael P. and Stephen C. Smith (2004). *Economic Development* (8th Edition), Pearson Education (Singapore) Pte. Ltd.

Thorat, V.A., S.N. Tilekar, J.S. Dhekale and H.K. Patil (2006). "Total Factor Productivity in Horticultural Crops in Konkan Region of Maharashtra", *Agricultural Economics Research Review* 19: 113-120.

Tripathi, A.K., S. Mandal, K.K. Datta and M.R. Verma (2006). A Study of Marketing of Ginger in Ri-Bhio District of Meghalaya", *Indian Journal of Agricultural Marketing* 20(2): 106-116.

United Nations Environment Programme (UNEP) (1992). "Global Biodiversity Strategy: Guidelines for Action to Save Study and Use Earth's Biotic Wealth Sustainably and Equitably", *Convention on Biodiversity*. Geneva: UNEP.

———. (1994). *Convention on Biological Diversity: Text and Annextures*. Switzerland: Interim Secretariat for the Convention on Biological Diversity, Geneva Executive Center.

———. (1995). *Global Biodiversity Assessment*. Cambridge: Cambridge University Press.

Upadhyay, R.C. and R.C. Sundriyal (1998). "Crop Gene Pools in the Northeast Indian Himalayas and Threats", in T. Partap and B. Sthapit (eds.), *Managing Agrobiodiversity: Farmers' Changing Perspectives and Institutional Responses in the Hindu Kush-Himalayan Region*.

Virchow, D. (1999). *Conservation of Genetic Resources: Costs and Implications for Sustainable Utilization of Plant Genetic Resources for Food and Agriculture.* New York: Springler Publicaton. pp.84-87.

Widawsky, D.A. and J.C. O'Toole (1996). "Prioritizing the Rice Research Agenda for Eastern India", in R.E. Evenson, R.W. Herdt and M. Hossain (eds.), *Rice Research in Asia: Progress and Priorities*. Wallingford, Oxford, UK: CAB International. pp.109-130.

Wood, D. (1988). "Crop Germplasm: Common Heritage or Farmers Heritage?", in R. Jack Kloppenburg (ed.), *Seeds and Sovereignty*. Duke University Press.